Computational Chemistry

Guy H. Grant

Lecturer in Biochemistry,
University College, Dublin

W. Graham Richards

Reader in Computational Chemistry,
Physical Chemistry Laboratory, University of Oxford

OXFORD   NEW YORK   TORONTO
OXFORD UNIVERSITY PRESS
1995

Oxford University Press, Walton Street, Oxford OX2 6DP

Oxford   New York
Athens   Auckland   Bangkok   Bombay
Calcutta   Cape Town   Des es Salaam   Delhi
Florence   Hong Kong   Istanbul   Karachi
Kuala Lumpur   Madras   Madrid   Melbourne
Mexico City   Nairobi   Paris   Singapore
Taipei   Tokyo   Toronto

and associated companies in
Berlin   Ibadan

Oxford is a trade mark of Oxford University Press

Published in the United States
by Oxford University Press Inc., New York

A catalogue record for this book is available from the British Library

Library of Congress Cataloging in Publication Data
Grant, Guy H.
Computational chemistry / Guy H. Grant, W. Graham Richards. — 1st ed.
(Oxford chemistry primers; 29)
1. Chemistry—Data processing.   2. Chemistry—Computer stimulation.
I. Richards, W. G. (William Graham)   II. Title.   III. Series.
QD39.3.E46G73   1995   542'.85—dc20   94-24733

ISBN   0-19-855740-X (Pbk)

Typeset by the authors and AMA Graphics Ltd., Preston, Lancs
Printed in Great Britain by
The Bath Press, Avon

Series Editor's Foreword

Oxford Chemistry Primers are designed to provide clear and concise introductions to a wide range of topics that may be encountered by chemistry students as they progress from the freshman stage through to graduation. The Physical Chemistry series will contain books easily recognized as relating to established fundamental core material that all chemists will need to know, as well as books reflecting new directions and research trends in the subject, thereby anticipating (and perhaps encouraging) the evolution of modern undergraduate courses.

In this Physical Chemistry Primer, Guy Grant and Graham Richards have produced a clearly written and fascinating introductory account of *Computational Chemistry*. This is an ever increasingly powerful subject which seeks to predict quantitatively molecular (and biomolecular) structure, properties, and reactivity by computational methods alone; it has enormous importance and potential in many areas of chemistry ranging from the fundamental to the applied. The Primer will be of interest to all students of chemistry and their mentors.

Richard G. Compton
Physical Chemistry Laboratory, University of Oxford

Preface

What is computational chemistry? In attempting to answer this question our aim has been to explain many of the underlying principles and assumptions of the subject while at the same time showing, through examples, how its growth has been driven by the problems it has been used to investigate. Many of these techniques are firmly rooted in established areas of physical chemistry.

Our starting point is the very detailed descriptions of molecules which come from quantum chemistry. Spectroscopists have simplified this picture by representing a molecule's motions in terms of classical force fields which, in turn, has led to the development of the molecular mechanics method. As one moves on to larger systems one must consider how molecules interact with each other, and it is here that experimental and theoretical studies of intermolecular forces have shown how more simple potential functions can be used to approximate molecular behaviour. Finally, simulations of small ensembles of molecules can be extrapolated to predict bulk behaviour, of the type measured by experiment, using the techniques of statistical mechanics. However, although much of the theory has been known for many years, it is only recent developments in computer hardware which have allowed application to large molecular systems.

The emphasis of the subject matter of this book is on biological systems. This has been quite deliberate in that it reflects the authors' research interests, but more importantly, it is here that computational chemistry has had the greatest impact, both academically and industrially, in providing detailed descriptions of molecular behaviour.

Dublin   G. H. G.
Oxford   W. G. R.
October 1994

Contents

1 Introduction

Chemistry is an experimental subject. That truism was self-evident historically, but during the past two decades it has ceased to be valid. In many areas of chemistry, theoretical prediction of chemical properties can rival experimental measurement. The choice of whether to find the structure of a molecule, its bond lengths and bond angles, by experimental spectroscopy or x-ray crystallography or by computation may come down to one of convenience or cost.

Cost is an important factor. Unlike in almost every other area of human existence, the costs of computing have fallen steadily over the years while at the same time the power in terms of hardware and software has grown and grown. This has encouraged first of all theoretical chemists and later physical, organic and inorganic chemists to use computational techniques. Frequently this is done by using standard packages which are akin to pieces of experimental apparatus. They may be treated almost like 'black boxes', as may NMR spectrometers. Like the latter, however, the best and most reliable results come from workers who understand how the 'black box', the computer program, works and the strengths and weaknesses of the techniques.

A sub-discipline of computational chemistry has grown up. This topic is distinct from theoretical chemistry. The activity involves taking known theory and developing the computer software to solve chemical problems. Very rapidly the areas of application have grown with, for example, many computational chemists being employed in industry, particularly the pharmaceutical industry.

1.1 Hardware

Computer hardware extends from personal computers, through workstations and mainframe machines to supercomputers and massively parallel devices. Computational chemistry is performed on all of these platforms with the obvious restriction that the bigger the problem, in terms of the size of molecule or the numbers of molecules involved, the more power or computer time required.

Although personal computers can be used to visualize molecules, displaying graphical representations, calculations beyond this level are somewhat restricted. Workstations which cost of the order of £10,000, by contrast can be used for almost the entire range of computational chemistry. Such machines have a large amount of memory and are often joined *via* ethernet links to other computers. An arrangement of this type is currently the favoured option of many practitioners, particularly those involved in industrial research. They may have as much power as the larger mainframe machines which are more likely to be found in central facilities, rather than on an individual chemist's desk. Workstations also have the attraction of not requiring air-conditioned surroundings or the presence of dedicated support staff.

Supercomputers have been defined as machines which cost more than $10 million and represent the top end of the scale of computational power. They are necessary for the biggest problems, particularly if many millions of integrals have to be performed or if the system being studied comprises many thousands of molecules. Computational chemists are amongst the largest users of supercomputer time and to serve their needs a number of the world's leading industrial companies have become the owners of the most powerful machines.

Computational chemistry is also seen as the application area for which many of the more novel types of computer architecture have been developed. Above all, these recent advances, such as the transputer, are designed to run many calculations in parallel rather than sequentially. A large part of the work within computational chemistry is ideally suited to such an approach.

1.2 Software

Many computational chemists are users of big packages; sophisticated programs developed to do a particular task. Most of these programs have been written in FORTRAN although graphical programs are increasingly written in 'C'. The programs are often very large, perhaps several hundred thousand lines of code. Clearly no single individual writes such a program. Rather they have grown as generations of users have added new functionality or expanded options. Great care has generally been taken to simplify input and output, particularly with the use of graphics. As a result it may be possible to do a great deal of computational chemistry without writing code. Computational chemists who like writing programs do occasionally write a novel program from scratch, but more typically they adapt or extend the capabilities of existing software. For a student of computational chemistry the progression is akin to becoming an expert with motor cars. First, one must learn how to run a big package – like learning to drive. Secondly, faults have to be detected and put right. As the student becomes more confident it becomes possible to modify the program so as to improve performance, perhaps cannibalizing bits from other programs; building special routines. Ultimately one can build one's own program, but using ready-made components in the form of sub-routines or sections of other people's code. It is virtually never necessary to start with a blank sheet of paper and compose a totally new program.

1.3 Areas of application

The range of applications of computational chemistry grows by the month, but the main classes of problem which can be resolved by computer are summarised here with further details given in later chapters.

Single molecule calculations

Calculations on a single molecule can give information about the most stable geometry of the molecule; its bond lengths and angles. Barriers to internal rotation round single bonds and conformational information can equally come from these calculations as can vibrational frequencies; electron distribution; ionization potentials; electron affinities; dipole moments; spin-orbit coupling constants and in principle any physical observable. In general such quantities can be computed to an accuracy comparable with experimental accuracy for molecules with up to about twenty atoms. Even excitation energies to excited electronic states can be calculated.

Assemblies of molecules

When large numbers of molecules, in particular solvent molecules, are incorporated in computations then thermodynamic properties can be calculated. Thus we can calculate not just the energy of an individual molecule but enthalpies and free energies. These lead to heat capacities and equilibrium constants. Equilibrium constants cover the whole range of equilibria, not just chemical equilibria. Thus binding energies between small molecules and macromolecules; partition coefficients and ionization constants are possible solvable problems as is the calculation of potentials of mean force or free energy as a function of some coordinate.

Reactions of molecules

Even computational chemistry cannot predict the rate constants for chemical reactions, which remains perhaps the single biggest unsolved problem in chemistry. However, computer calculations can give an idea of relative rate constants and structures of transition states. It has been possible to include solvent effects in simple reactions and to follow the behaviour of macromolecules over periods of perhaps a few hundreds of picoseconds (1 ps = 10^{-12} seconds).

1.4 Computational developments

Every time a significant advance has been made in computing technology or in application then computational chemists have seized the opportunity and incorporated it into their own field. Two examples will suffice to illustrate.

Computer graphics

Perhaps of all advances in the last decade or so the impact of computer graphics has had the biggest influence on computational chemistry. Graphics provided the link between computer professionals and traditional laboratory bench chemists. Chemists have always used molecular models to understand their subject. However as the molecular systems studied become larger and larger the drawbacks of mechanical models became apparent. It may take weeks or months to build a model of a protein from the x-ray crystal coordinates of atom position. The resulting model will be very crude since all bond lengths and angles will be standard values chosen for building blocks; the model will suffer from gravity – bits fall off, and it is difficult to see or measure distances in the interior of the model. Computer-graphic displays overcome all these disadvantages and can utilize the full range of display used in computer games. Colour; three-dimensional viewing; motion of atoms; the ability to rotate, cut away or highlight features. Even the techniques of virtual reality are already making an impact in molecular computer graphics. Beyond merely displaying models of molecules, graphics terminals and workstations can be used to reveal molecular properties such as electron densities or the energy of interaction of point charges with a molecule, to give some indication of reactivities or intermolecular binding preferences.

Neural networks

At its crudest level the human brain can be considered as a set of interconnecting cells or neurons with the process of learning being based on developing a range of connections of variable strength between these elements. This essential idea is the basis of the neural network. The network has a number of processing elements or nodes with connections between them. Each element takes in a signal from elements to which it is connected in the previous layer, and sends out a signal to the next layer. The strength of signal being passed depends on the 'weight' of each of the connections, and it is these weights which are varied. First, a set of input data is connected by the network to an output. When both input and output are known the weights are adjusted during 'training' so that the output is appropriate to the input. For example, a set of molecular properties may be used as input to predict another property, perhaps an ionization potential. When the net is trained to give the right answer for a training set of molecules it can then be tried on cases where the input properties are known but not the output. This technique is in many respects a form of pattern recognition and is becoming widespread as a way of predicting values of properties which are hard to measure accurately particularly in biochemical applications.

1.5 Summary

Computational chemistry is one of the fastest growing areas of chemistry. Although there are specialists in the field, increasingly the techniques are being applied by experimental chemists using the ever-growing power of ever-cheaper computers. A knowledge of the topic is important for anyone contemplating chemical research.

2 Quantum mechanics

2.1 Wave functions

The Schrödinger equation lies at the heart of much of modern science. In its barest form it states

$$H\psi = E\psi$$

Here H is the shorthand notation for an operator which operates on a mathematical function, the wave function ψ, and E is the energy of the system. This notation disguises the fact that this equation is a differential equation, or rather a set of equations, with a function ψ_n corresponding to each allowed energy E_n.

In the case of the simple hydrogen atom system with a single electron outside a positively charged nucleus, the equation may be solved exactly providing the wave function, ψ, obeys a set of reasonable restrictions on its behaviour.

It must be emphasized that ψ is just a normal mathematical function. It has no supernatural properties. Plotting a function $f(x)$ against a coordinate x is a familiar procedure (Fig. 2.1). In molecular quantum mechanics, since the problems concern three-dimensional molecular systems, then the wave function also varies in these coordinates. Because of the essentially spherical nature of atoms, polar coordinates (r, θ and ϕ) are generally preferred over Cartesian, x, y, z, coordinates (Fig. 2.2)

For the hydrogen atom the allowed wave functions or eigenfunctions are well known and an attempt to represent the three-dimensional functions in the two dimensions of the printed page is given in Fig. 2.3.

It can be seen in the figure that these resulting functions, one for each allowed energy or eigenstate, do satisfy the postulates of quantum mechanics in that they are 'well-behaved' functions. They go to zero at infinity and change smoothly, never doubling back on themselves, having discontinuities or even violent changes in curvature.

Once the wave function is known for a particular state of a system then any physical observable may in principle be determined using the prescription

$$\text{Observable} = \frac{\int \psi^*(\text{operator})\psi \, d\tau}{\int \psi^*\psi \, d\tau}$$

Integration over the element $d\tau$ means over all space and ψ^* is the complex conjugate of ψ; ($\psi^*\psi$) being the modulus of the square of the wave function. There is an appropriate operator for each different observable, be it

Fig. 2.1 Graph of the function $f(x)$ against the variable x.

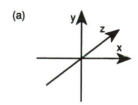

(a)

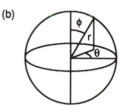

(b)

Fig. 2.2 (a) Cartesian, x, y and z, coordinates. (b) Polar, r, θ and ϕ, coordinates .

energy when the hamiltonian operator, H, is the one to use, or another, for the charge density, dipole moment or hyperfine coupling constant.

2.2 Orbitals

The wave functions which satisfy the Schrödinger equation for the hydrogen atom are sometimes called *orbitals*. A hydrogenic atomic orbital is thus merely a three-dimensional mathematical function from which one can calculate the energy or other properties of the single electron system.

In polyelectronic atoms we adopt the so-called *orbital approximation*. This involves treating each electron separately, each with its own one-electron wave function or orbital. This mathematical approximation is nothing more than the fundamental basis of the universal procedure of describing atoms by means of orbital configurations. Thus to write atomic electron configurations such as

Li : $1s^2 2s$

or

C : $1s^2 2s^2 2p^2$

is actually a mathematical approximation which treats each electron separately. In lithium two electrons have functions associated with them which are of the $1s$ shape and one with $2s$ form.

An orbital is thus merely a synonym for a one-electron wave function. Each is a three-dimensional mathematical function which describes the behaviour of a single electron.

Generalizing this, with some formal notation, we may write for a poly-electronic atomic system, the total wave function for an atom, ψ, is a product of one-electron atomic wave functions (χ_i), one for each electron, i.e.

$$\psi = \chi_1 \chi_2 \chi_3 \cdots \chi_n$$

2.3 Spin-orbitals and antisymmetry

When we write the electronic structure of beryllium as $1s^2 2s^2$ we understand this as being a shortened form of

$$1s^\alpha \, 1s^\beta \, 2s^\alpha \, 2s^\beta$$

where α and β represent the two opposite directions of electron spin.

If we include the spin in our atomic orbitals writing either χ^α or χ^β or the equivalent shorthand, χ for an orbital associated with α spin and $\bar{\chi}$ for one with β spin, then our wave function for the whole atom is a product of *spin-orbitals*, for example:

He: $1s(1) \, 1\bar{s}(2)$,

and

Be: $1s(1) \, 1\bar{s}(2) \, 2s(3) \, 2\bar{s}(4)$.

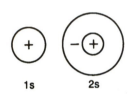

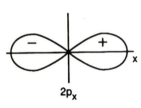

Fig. 2.3 Allowed wave functions for the hydrogen atom.

In this expanded form of the common shorthand the numbers in parentheses indicate which electron is associated with each spin-orbital.

The expanded form of the notation enables us to see that we will need to make things a little more complicated before simplifying once more by means of a convenient notation.

Let us consider the case of the helium atom. If we write

$$\psi_{He} = \chi_{1s}^{\alpha}(1)\chi_{1s}^{\beta}(2)$$

or alternatively and equivalently

$$\psi_{He} = 1s(1)\,1\bar{s}(2)$$

we can see that we are not satisfying the Pauli principle. In its fundamental form this states that the wave function for the system (ψ) must change sign if we interchange any pair of electrons, since electrons are identical fermion particles.

We have written our wave function as a simple product of orbitals when this is not sufficient, since

$$\psi = 1s(1)\,1\bar{s}(2)$$

but on changing we get

$$\psi' = 1s(2)\,1\bar{s}(1)$$

and ψ is not the negative of ψ'.

To overcome this we can write the wave function for the atom in the orbital approximation as

$$\psi = \frac{1}{\sqrt{2}}\big[1s(1)\,1\bar{s}(2) - 1s(2)\,1\bar{s}(1)\big]$$

Now if we make the interchange, Pauli's principle will be satisfied (and the $1/\sqrt{2}$ retains the normalizing condition, $\psi^{*}\psi\,d\tau = 1$). Conventionally we write a volume element in three dimensions as dv, but if spin is also included the element is written as $d\tau$.

For beryllium, however, we must allow for any possible interchanges between four electrons so then

$$\begin{aligned}
\psi = \ & 1s(1)\,1\bar{s}(2)\,2s(3)\,2\bar{s}(4) - 1\bar{s}(1)\,1s(2)\,2s(3)\,2\bar{s}(4) + \\
& 2s(1)\,1\bar{s}(2)\,1\bar{s}(3)\,2s(4) - 2\bar{s}(1)\,1s(2)\,1\bar{s}(3)\,2s(4) + \ldots \text{ etc.} \\
& \text{in all 24 such products.}
\end{aligned}$$

Fortunately notation can help us. All these products are simply the expanded form of determinants.

For He

$$1s(1)\,1\bar{s}(2) - 1s(2)\,1\bar{s}(1)$$

is the expansion of

$$\begin{vmatrix} 1s(1) & 1s(2) \\ 1\bar{s}(1) & 1\bar{s}(2) \end{vmatrix}$$

which is scarcely shorter. But for Be our long expansion above can be summarized as

$$\psi = \frac{1}{\sqrt{4!}} \begin{vmatrix} 1s(1) & 1s(2) & 1s(3) & 1s(4) \\ 1\bar{s}(1) & 1\bar{s}(2) & 1\bar{s}(3) & 1\bar{s}(4) \\ 2s(1) & 2s(2) & 2s(3) & 2s(4) \\ 2\bar{s}(1) & 2\bar{s}(2) & 2\bar{s}(3) & 2\bar{s}(4) \end{vmatrix}$$

Since the diagonals of these determinants are sufficient to define them and because the value of the normalization constant is obvious ($1/\sqrt{n!}$ where n is the number of electrons), these expanded spin-orbital wave functions are usually written as simple products

$$\psi_{He} = 1s1\bar{s} \text{ or even } \psi_{He} = 1s^2$$
$$\psi_{Li} = 1s^2 2s$$
$$\psi_{Be} = 1s^2 2s^2$$

The fact that these expressions are shorthand notation for determinantal wave functions is hidden in the notation.

When we want the wave function of an atom, ψ, we write it in the form

$$\psi = \chi_1 \chi_2 \chi_3 \cdots \chi_n$$

remembering that this is not a simple product.

The problem then becomes one of finding χ_1 etc. Each of these is a one-electron function in coordinates, r, θ and ϕ multiplied by a spin factor α or β. The individual χ_i can be taken as being hydrogen-like analytical functions but with different exponents appropriate to the particular atom or alternatively expressed (as any function can be) as a numerical value for each point defined by the three coordinates. This latter type of atomic orbital is called a numerical function.

Hartree produced some extremely accurate atomic functions of this numerical type which were later fitted to analytic forms by Slater and are called Slater atomic wave functions. They have the form

$$\chi_i = \text{Normalization constant} \times (\text{exponential function of } r)$$
$$\times (\text{spherical harmonic in terms of } \theta \text{ and } \phi)$$

The spherical harmonic part is identical to the shape variation found for hydrogen atom wave functions and the differences from atom to atom are found only in the r-dependent, or radial, part of the orbital. The quality of a wave function may be tested by using it to compute the energy of the system and using the well-known variation principle which states that the better the wave function the lower will be the resulting energy.

The atomic orbitals, χ_i, for all atoms encountered in chemical systems are well known and need only be taken from the literature.

A useful property of orbital functions of a given atom is that they may be made orthonormal. This means

$$\int \chi_i^2 \, d\tau = 1$$

integrating over space and spin coordinates, but

$$\int \chi_i \chi_j \, d\tau = 0 \text{ if } i \neq j$$

Generally we use wave functions which are normalized but not orthogonal, i.e.

$$\int \chi_i \chi_j \, d\tau = S_{ij}$$

This is the definition of the overlap integral S_{ij}.

2.4 Molecular orbitals

The wave function for a molecule is not in principle any different from that for an atom. We may use the symbol Ψ, to represent the molecular wave function and, as in the atomic case, we can make the orbital approximation,

$$\Psi = \phi_1 \phi_2 \phi_3 \ldots \phi_n$$

where each function ϕ_i will be a three-dimensional function which determines the properties of an individual electron in the molecule. We may include spin so that the wave function is a product of spin-orbitals. We must also remember that the wave function has to be antisymmetric with respect to electron interchange, or renumbering the electrons, with the result that the product as written represents the diagonal of a determinant.

The goal of most quantum molecular calculations is the production of a molecular wave function Ψ. This will be achieved if we know all the constituent molecular orbitals ϕ_i. In most of the methods currently applied to molecular questions the problem is broken down one stage further. We expand each of the unknown molecular orbitals ϕ_i as a linear combination of the known atomic orbital functions.

Thus

$$\phi_i = \sum_k c_{ik} \chi_k$$

Each of the χ_k will be the function of the form

$$\chi_k = \text{Constant} \times (\text{function of } r)$$
$$\times (\text{spherical harmonic function in terms of } \theta \text{ and } \phi)$$

Here i is an index which labels the particular molecular orbital and k is a running index 1, 2, 3 . . . whose range will depend on just how big an expansion is taken. Our problem in finding the wave function for the molecule, which means calculating the molecular orbitals ϕ, now reduces to finding the expansion coefficients c_{ik}.

For the simple case of a diatomic molecule like Li_2 this process can be illustrated pictorially by means of a molecular orbital diagram (Fig. 2.4), in which we represent each molecular orbital as a sum of two lithium atomic orbitals. The molecular orbitals are labelled $1\sigma_g$, $1\sigma_u$ and $2\sigma_g$ for reasons of symmetry.

The molecular wave function for the ground state of Li_2 where the lowest (most tightly bound) molecular orbitals are each doubly filled will thus be

$$\Psi \ (Li_2; \text{ground state}) = \phi_{1\sigma g} \ \phi_{1\sigma g} \ \phi_{1\sigma u} \ \phi_{1\sigma u} \ \phi_{2\sigma g} \ \phi_{2\sigma g}$$

with

$$\phi_{1\sigma g} = c_{11}\chi_{1sLi} + c_{12}\chi_{1sLi}$$

and

$$c_{11} = c_{12} \text{ by symmetry}$$

$$\phi_{1\sigma u} = c_{21}\chi_{1sLi} + c_{22}\chi_{1sLi}$$

and

$$c_{21} = -c_{22}$$

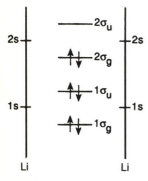

Fig. 2.4 Molecular orbitals for diatomic Li_2.

The use of symmetry is convenient in small symmetrical molecules but plays very little part in molecules of biological interest which only rarely have symmetry.

An important principle in quantum mechanics, the variation principle, states that the more flexible the wave function the lower the energy and by implication the better the wave function. Zero on the energy scale has all particles separated at infinity. Calculated energies are thus negative numbers, the bigger the number the 'lower' the energy.

In the Li_2 example we can get a 'better' or 'lower' energy by using a more flexible wave function, perhaps extending our molecular orbital expansion

$$\phi_i = c_{i1}\chi_{1sLi} + c_{i2}\chi_{2sLi} + c_{i3}\chi_{2pLi}$$

However lengthy our expansion the problem will remain the same: to obtain the molecular wave function, Ψ, we find the molecular orbital ϕ in terms of known functions multiplied by coefficients which have to be determined. This is done by the solution of the secular equations.

2.5 Secular equations

We have met the shorthand form of the Schrödinger equation

$$H\psi_n = E_n\psi_n$$

A similar equation may be written for each of the molecular orbitals

$$H\phi_i = \varepsilon_i\phi_i$$

but now H is a one-electron hamiltonian, whose specific nature we will discuss later, and ε_i is the orbital energy or energy of one particular electron in orbital i. However we have seen that ϕ may be expanded as a linear combination of known atomic orbitals. Hence the one-electron equations may be rewritten as

$$H\sum_k c_{ik}\chi_k = \varepsilon_i\sum_k c_{ik}\chi_k$$

Now we may multiply each side of the equation by χ_i (any one of the normalized set which includes χ_k) and integrate over all the electronic space coordinates (dv) giving the secular equations

$$\sum_k c_{ik}\left(\int \chi_l H\chi_k \, dv\right) = \varepsilon_i\sum_k \left(\int \chi_l\chi_k \, dv\right)$$

Conventional notation writes

$$H_{lk} = \int \chi_l H\chi_k \, dv$$

and

$$S_{lk} = \int \chi_l\chi_k \, dv$$

By this means the equations may be tidied up into the form

$$c_{ik}\left(H_{lk} - \varepsilon_i S_{lk}\right) = 0$$

Such a set of equations will only have a non-trivial solution if the following conditions hold

$$\det\left|H_{lk} - \varepsilon S_{lk}\right| = 0$$

This simple determinantal equation is the basis of all molecular orbital methods. In general the methods go directly to the secular determinant. All the terms (or matrix elements) H_{lk} and S_{lk} are computed. If the determinant were multiplied out this would yield a polynomial in ε, solutions of which would give the allowed molecular eigenenergies or orbital energies. Each ε_i in turn is then put into the secular equations and hence the desired coefficients are found.

Example

Let us consider a rather formal example which will illustrate the principles involved in calculating molecular orbital coefficients and is closely related to some genuine applications.

Suppose we are expanding the molecular orbitals in terms of three atomic orbitals

$$\phi_i = c_{i1}\chi_1 + c_{i2}\chi_2 + c_{i3}\chi_3$$

The one-electron equation is

$$H\phi_i = \varepsilon\phi_i$$

so that we have three such equations and can determine three molecular orbitals. We now multiply both sides of the equation successively by χ_1, χ_2 and χ_3 and integrate, giving three secular equations

$$c_{11}\,(H_{11} - \varepsilon S_{11}) + c_{12}\,(H_{12} - \varepsilon S_{12}) + c_{13}\,(H_{13} - \varepsilon S_{13}) = 0$$
$$c_{21}\,(H_{21} - \varepsilon S_{21}) + c_{22}\,(H_{22} - \varepsilon S_{22}) + c_{23}\,(H_{23} - \varepsilon S_{23}) = 0$$
$$c_{31}\,(H_{31} - \varepsilon S_{31}) + c_{32}\,(H_{32} - \varepsilon S_{32}) + c_{33}\,(H_{33} - \varepsilon S_{33}) = 0$$

For solution we must solve the determinantal equation

$$\det \begin{vmatrix} H_{11} - \varepsilon S_{11} & H_{12} - \varepsilon S_{12} & H_{13} - \varepsilon S_{13} \\ H_{21} - \varepsilon S_{21} & H_{22} - \varepsilon S_{22} & H_{23} - \varepsilon S_{23} \\ H_{31} - \varepsilon S_{31} & H_{32} - \varepsilon S_{32} & H_{33} - \varepsilon S_{33} \end{vmatrix} = 0$$

In this case we have a cubic equation in ε which has three roots.

Since more usually our expansions go well beyond three terms a more convenient method of solution of the determinantal equation is employed which makes use of the strength of computers.

Rows and columns of determinants may be added and subtracted or factors divided out without altering the value of the determinant. If by such adjustments we could rearrange the determinant into the form

$$\begin{vmatrix} (x - \varepsilon) & 0 & 0 \\ 0 & (y - \varepsilon) & 0 \\ 0 & 0 & (z - \varepsilon) \end{vmatrix} = 0$$

where x, y, and z are numbers and all the rest of the determinant is zero, on multiplying out we should have the simple equation

$$(x - \varepsilon)(y - \varepsilon)(z - \varepsilon) = 0$$

Computational methods for doing this or nearly equivalent simplifications are standard procedures. As a result there is no difficulty in solving the determinantal equation and hence finding the molecular orbital coefficients, no matter how extensive the expansion. The actual computer programs are invariably based on matrix diagonalization techniques which are equivalent to the above description, but particularly suitable for computers.

2.6 Matrix elements

Going directly to the secular determinant and thence to the secular equations to obtain molecular orbitals poses no serious problem providing all the 'matrix elements' H_{lk} and S_{lk} are easily found.

The matrix elements S_{lk} are called overlap integrals as they represent the overlap between the two three-dimensional functions χ_l and χ_k

$$S_{lk} = \int \chi_l \chi_k \, dv$$

These integrals are reasonably easy to evaluate and again standard computer programs are available.

The matrix elements H_{lk} or

$$\int \chi_l H \chi_k \, dv$$

on the other hand include the one-electron operator H. So far the precise nature of H has not been specified and indeed it is in the definition, or lack of it, that most of the molecular orbital methods differ. In the more sophisticated methods H is precisely defined. In the less precise approximations it is never defined and all H_{lk} matrix elements are replaced by parameters.

2.7 Self-consistent molecular orbitals

The most clearly defined molecular orbital calculations are based on the Hartree-Fock method. A hamiltonian operator contains terms for the kinetic and potential energy of the system. A suitable choice of units makes the equations look rather tidier than standard units. If we choose to take the electronic charge, e, and mass, m_e, each as unity and the unit of length as the Bohr radius then the Schrödinger equation for the hydrogen atom becomes

$$\left(-\frac{1}{2}\nabla^2 - \frac{1}{r} \right)\Psi = E\Psi$$

where the kinetic energy is represented by the $-1/2\nabla_i^2$ term (∇^2 is shorthand for $d^2/dx^2 + d^2/dy^2 + d^2/dz^2$) and potential energy by the electron-nuclear attraction, $1/r$.

For the hydrogen molecule there are two electrons and the Schrödinger equation for the molecular wave function Ψ is

$$\left\{ -\frac{1}{2}\nabla_1^2 - \frac{1}{2}\nabla_2^2 - \frac{1}{r_{1A}} - \frac{1}{r_{1B}} - \frac{1}{r_{2A}} - \frac{1}{r_{2B}} + \frac{1}{r_{12}} \right\}\Psi = E_{el}\Psi$$

To this electronic energy E_{el} has to be added the nuclear-nuclear repulsion, $E_{nuclear} = 1/R$, where R is the separation of the nuclei A and B.

It may be seen that this equation represents the sum of two separate hydrogen molecule ion wave equations with the additional term, $1/r_{12}$, representing the repulsion between the two electrons.

When we come to one-electron equations we want to include in the hamiltonian terms for all the energy contributions of that one electron. These will be kinetic energy, nuclear attraction and electron-electron repulsion. The kinetic energy term in the operator is $-1/2\nabla_i^2$ and the various nuclear attraction terms are

$$\sum_v Z_v / r_{iv}$$

where the nuclei are labelled by v and the particular electron we are considering is electron i. These two parts can be grouped together and designated as H^N. If there were no other electrons in the molecule this would be a sufficient hamiltonian and the one-electron energy would be the full molecular electronic energy which we could call ε_i^N.

When, as is usually the case, there are many electrons in the molecule a major part of the potential energy which must be represented in the hamiltonian is electron-electron repulsion. Now if we have two electrons with orbitals ϕ_i and ϕ_j, separated by a distance r_{12}, then the repulsion between them is given by

$$\int \phi_i^2(1)\, \frac{1}{r_{12}}\, \phi_j^2(2)\, dv_1 dv_2$$

since

$$\int \phi_i^2(1)\, dv_1$$

is the charge distribution of electron (1) (assuming ϕ_i is real). If ϕ is complex this should be replaced by

$$\int \phi_i^*(1)\phi_j(1)\, dv_1$$

The expression

$$\int \phi_j^2(2)\, dv_2$$

is the charge distribution for electron (2).

Thus we must include terms of this type in the one-electron hamiltonian, suggesting that an expanded form of

$$H\phi = \varepsilon\phi$$

might be

$$\left\{ -\frac{1}{2}\nabla^2 - \sum_v \frac{Z_v}{r_{iv}} + \sum_{j=1}^{n} \phi_j^2(2)\frac{1}{r_{12}}\, dv_2 \right\} \phi_i(1) = \varepsilon_i\phi_i(1)$$

This equation due to Hartree would be correct if our orbital wave function were a simple product and not a determinant. To account for the determinantal form of Ψ the full self-consistent field equations or Hartree-Fock equations are

$$H^{SCF}\phi_i = \varepsilon_i\phi_i$$

where

$$H^{SCF} = \left\{ H^N + \sum_j J - \sum_j{}' K \right\}$$

and the shorthand notation used is defined by

$$J_j\phi_i(1) = \left(\int\phi_j^2(2)\,\frac{1}{r_{12}}\,dv_2 \right)\phi_i(1)$$

and

$$K_j\phi_i(1) = \left(\int \phi_j(1)\,\phi_i(2)\frac{1}{r_{12}}\,dv_2 \right)$$

with a prime on a summation indicating summing only over pairs of electrons of the same spin.

The Hartree-Fock equations contain the K_j or exchange terms in addition to the obvious Coulombic interelectron interactions allowed for in the Hartree equations. These exchange terms arise as a result of the Pauli principle which leads to determinantal wave functions rather than products and exchange integrals between various cross-products of the expanded determinant.

This specification of H in the one-electron equation leads to a lot of terms, but that is not a particularly difficult problem if we are using a computer.

More seriously the hamiltonian in $H\phi = \varepsilon\phi$ itself contains the various ϕ which we are trying to determine. That is to say the coefficients in

$$\phi_i = \sum c_{ik}\chi_k$$

must be known before we start. This is the origin of the term 'self-consistent' field. Starting values of c_{ik} are given from educated guesses or from the results of simple calculations. The determinantal equation

$$\det\left| H^{SCF} - \varepsilon S_{lk} \right| = 0$$

is solved after calculating all the integrals involved in H_{lk} and S_{lk}. Solution yields values of ε which are substituted in the secular equations to give new values of the various c_{ik}. The process may then be repeated until the c_{ik} resulting from one cycle are identical within prescribed limits with those used in the previous cycle. The results are then self-consistent.

The various simplified methods of calculating molecular orbitals are essentially approximations of a greater or less drastic nature which result in a reduction of the number of integrals necessary to build the matrix element H_{lk} and S_{lk} in the determinantal equation.

2.8 Configuration interaction

We have already met the variation principle which tells us that the more flexible a wave function the better it will be in terms of the energy which results. One way of improving the wave functions which we have as a result of self-consistent field or approximate calculation is to allow the interaction of configurations.

Formally this means allowing a further linear mixing to give an improved wave function,

$$\Psi_{improved} = a\Psi_o + b\Psi_1 + c\Psi_2 + \dots$$

In this expression Ψ_o is the wave function we first find and Ψ_1, Ψ_2 etc. are wave functions which would be appropriate for excited configurations of the same symmetry. The coefficients a, b, c etc. are mixing coefficients whose values are chosen so that energy improvement is maximized.

The problem with configuration interaction is knowing when to stop. Every extra configuration helps the energy a little. Although configurations closest in energy to the ground state may have an obvious influence others which are highly excited may also be significant.

In many molecular orbital packages the effects of excited states are incorporated using perturbation theory as introduced by Møller and Plesset.

2.9 *Ab initio* methods

All molecular wave functions are approximate; some are just more approximate than others. We can solve the Schrödinger equation exactly for the hydrogen atom but not even, despite what many textbooks say, for the hydrogen molecule ion, H_2^+. For H_2^+ we make the Born-Oppenheimer approximation which separates electronic and nuclear motion, and calculate the electronic energy of the ion with a given fixed internuclear distance and then obtain the total energy by adding the nuclear-nuclear repulsion term.

The term *ab initio* is perhaps unfortunate since it gives a spurious idea of quality, but it is universally used for calculations of orbital wave functions where the full Hartree-Fock self-consistent field operator is used in

$$\det \left| H_{lk} - \varepsilon S_{lk} \right| = 0$$

$$H = \left[H^N + \sum_j J - \sum_j{}' K \right]$$

and all the integrals implied in H_{lk} and S_{lk} are computed.

Each molecular orbital will be in the form

$$\phi_i = \sum c_{ik}\chi_k$$

If the expansion is infinite then we would achieve the most flexible wave function within the constraints of the self-consistent field hamiltonian which we have defined. The resulting energy would be the 'best' or biggest negative number we could obtain and is the Hartree-Fock limiting energy. In practice, if an expansion of thirty or forty terms is used for ϕ, then little further improvement in energy results and we can safely assume that we are close to the limit.

The larger the expansion in terms of atomic orbitals, the more integrals have to be computed and the more expensive computer time required. Thus for large molecules rather short expansions are used so that the resulting energy may be far from the best possible limit, and the wave function even though of *ab initio* origin is approximate even within the constraints of the method.

The most obvious set of atomic orbitals, χ_k, to use in the expansion are Slater-type atomic orbitals which were found by fitting analytical exponential functions to numerical atomic wave functions. If we use one such function for each atomic orbital which is filled, e.g. for C employing $1s$, $2s$ and $2p$ atomic orbitals, then the set of atomic functions or *basis set* is referred to as a *minimal basis*. If we use double that number, the basis set is of *double zeta* quality. Each of the atomic orbitals is of the form

$$\chi_k = Ce^{-\zeta r}Y_{lm}$$

Here C is a normalizing constant; Y_{lm} is the angular part of the function (a spherical harmonic), and ζ is called the orbital exponent, zeta. The Y_{lm} are well known for $1s$, $2s$, $2p$ etc. so that a basis set may be specified by listing the exponents for each type of orbital used. Suitable basis sets are available in the literature.

To overcome some of the problems in doing integrals with exponential (or Slater-type) wave functions, the more tractable gaussian form of orbital dependence is often employed. The gaussian radial dependence has the less appropriate $\exp(-ar^2)$ form. To combine the suitability of exponential basis functions and the simplicity of calculating with gaussians, the obvious step of fitting gaussian shapes to an exponential has been taken. Thus one frequently sees in the literature expressions such as 'calculations performed with an STO-3G basis'. This indicates that a minimal basis set of Slater exponentials has been used but for the integrals each exponential function was fitted by three gaussian functions. An STO-4G calculation is likely to be closer to a true minimal basis calculation but even that would be some way from the limit of the method in terms of energy.

When running an *ab initio* calculation the starting point is a particular molecular geometry, the nature and coordinates of each atom being defined. Depending on which atoms are in the molecule a basis set of atomic orbitals is then decided upon, although the choice may be built into the computer program. The program will then compute all the integrals required in building up H_{lk} and S_{lk} using guessed trial coefficients, build and diagonalize the determinant and produce a set of orbital energies and first-improved coefficients. As described earlier this process is repeated until self-consistency is achieved, when the program will print out a set of molecular orbitals ϕ_i in the form of the coefficients, and associated with each an orbital energy ε_i.

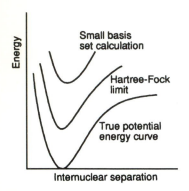

Fig. 2.5 Errors in molecular orbital calculations for a diatomic molecule as a function of internuclear separation.

It is important to realize that even if we can afford to have long expansions of molecular orbitals and even reach the Hartree-Fock limit, there are still defects in the wave functions which arise from approximations in the actual Hartree-Fock equations.

There are two sources of error in the starting equations. The first comes about because the whole theory is based on the Schrödinger equation which is not relativistically correct. Fast-moving inner electrons may move with speeds which are not negligible by comparison with the velocity of light and relativistic effects thus contribute; mass is not constant. Since most chemical and biological transformations of molecules do not involve core electrons this error is normally a constant and causes no serious difficulty.

The second error is more serious and is called the correlation energy error.

2.10 Correlation energy

Any defect in our wave functions will result in the calculated energy being less than the true energy. The correlation energy can be defined as

$$E_{\text{correlation}} = E_{\text{true}} - E_{\text{Hartree-Fock}} - E_{\text{relativistic}}$$

and represents the remaining energy error between the limiting Hartree-Fock energy and true total energy taking into account the relativistic effect. Correlation energy errors arise because of an unjustified approximation in the self-consistent field hamiltonian. This we write as

$$H = \left[H^{\text{N}} + \sum_{\text{j}} J_{\text{i}} - \sum_{\text{j}}' K_{\text{j}} \right]$$

Involved in J and K are terms of the type

$$\int \phi_{\text{j}}^2 \frac{1}{r_{12}} \, dv_2$$

These are included to account for the interaction of one electron with another, the second electron being represented as a smoothed-out averaged electron density. Thus if we are considering the helium atom, each of the $1s$ electrons, as far as the calculation is concerned, would interact with the other as if the second electron was spherically distributed. In reality, of course, the two electrons will have their positions correlated, there being a higher probability of the two electrons being on opposite sides of the nucleus than there is of them being both on the same side, as self-consistent field theory allows.

The correlation energy error results from electron pair effects and is reasonably constant as molecular geometry changes providing the electron pairings in the molecule do not change. It is not a constant as bonds are stretched to the point of dissociation (Fig. 2.5). Errors are thus quite serious in calculations of dissociation energies, and also in the estimations of ionization potentials or even pK_{a} values. On the other hand, if we merely change conformation and leave the electron pairings essentially unaltered, then the error may be constant (Fig. 2.6).

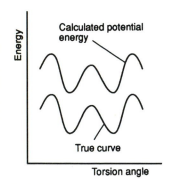

Fig. 2.6 Errors in molecular orbital calculations as a function of conformational change.

2.11 Semi-empirical calculations

Ab initio calculations are themselves not always perfectly successful in reproducing experimental observations. In addition the number of integrals required increases roughly as the fourth power of the number of basis functions in the molecule and for even quite small molecules many millions may be required. In consequence a great deal of effort has been expended devising so-called semi-empirical molecular orbital methods. These all start with the determinantal equation but make a variety of approximations to reduce the amount of computer time required.

All the commonly used techniques are valence electron calculations; that is they neglect $1s$ electrons since these play little part in chemical or biochemical behaviour. The $1s$ electrons are defined as part of the 'core' for first-row atoms and both **K** and **L** shell electrons for heavier atoms. The self-consistent field hamiltonian thus becomes

$$H = H^{core} + \sum_j J_i - \sum_j{}' K_j$$

and H^{core} incorporates kinetic energy and attraction to a core rather than to a bare nucleus. Integrals (or synonymously matrix elements) involving H^{core} are usually replaced by empirical or calculated parameters.

Particularly in instances where molecules of interest contain a heavy atom (such as a metal in an enzyme-binding site) an approximation at this level is introduced by incorporating a pseudopotential or effective potential into the hamiltonian. The resulting computed properties are not significantly different from those of full *ab initio* calculations in the cases of electron density or orbital energies. Savings of computer time by 50 per cent or more are however common.

There are two broad philosophies about the use of parameters in molecular orbital calculations. One school of thought takes the view that as *ab initio* calculations themselves are far from perfect, empirical parameters should be introduced so as to ensure agreement with experiment. In some ill-defined way this means that the parameterization includes correlation effects. Unfortunately correlation energy cannot be treated in this simple way in a universal fashion. Thus as long as the parameterization is related to experimental data which are closely related to those which one is trying to calculate, then this approach works perfectly. In the limit, of course, if you put some experimental data into the calculation you should at very least be able to get them out again. Generally semi-empirical methods using this philosophy give good agreement with experiment providing the experiment and the source of parameterization are related.

Alternatively there are methods which use parameters based on *ab initio* calculations and attempt to reproduce what a more costly rigorous calculation would have produced. This approach has the advantage of being clearly defined and it is not essentially biased in its parameterization towards one type of experiment or another. On the other hand the results can never be better than *ab initio* calculations.

The actual choice of method to be used in any research project will depend on the computer time available, the number of calculations required and above all on the agreement of test calculations of properties which are related to the problem with physical measurements.

2.12 Neglect of differential overlap

The matrix elements H_{lk} in the secular determinant involve a large number of integrals over atomic orbital functions χ of the type

$$\int \chi_m(1)\chi_n(1)\frac{1}{r_{12}}\chi_l(2)\chi_S(2)\,d\tau_1 d\tau_2$$

which are often written in the shorthand form (mn|ls). These integrals are particularly difficult to evaluate if the atomic functions, χ_m etc., are centred on different atoms. In particular those integrals involving three or four centres are time consuming, perhaps prohibitively so, even using a computer.

An approximation which will surmount this problem at a stroke is to assume

$$(mn|ls) = \delta_{mn}\delta_{ls}(mm|ll)$$

with δ_{mm} and δ_{ls} being Kronecker deltas which are equal to zero unless the subscripts are equal. In this way, neglecting differential overlap of functions based on different centres, we eliminate not only all three- and four-centre integrals but most two- and one-centre integrals where different atomic orbitals are involved for either of the two electrons.

The study of *ab initio* results where these neglected integrals are computed confirms that they are significantly smaller than the integrals which remain but they may be by no means negligible. Consequently further approximations and parameterizations are normally added to counteract the omissions.

In the CNDO, or complete neglect of differential overlap method, this approximation is fully applied for the valence electrons with $1s$ electrons being treated as part of a nuclear 'core'. Integrals which remain are further approximated. The electron repulsion integral (mm|ll) is supposed to depend only on the atoms A and B on which χ_m and χ_l are situated and set equal to the parameter A_B which may be found from an actual calculation for a simple example such as $(2s_A 2s_A | 2s_B 2s_B)$ or by the use of empirical data. Matrix elements of H^{core} are also parameterized and in particular, atomic ionization potentials are frequently used to replace integrals which approximately represent the energy with which an electron is held by an atom. There are a number of alternative parameterizations, details of which can be found in the references.

Slightly less dramatic in the application of the neglect of differential overlap are intermediate neglect of differential overlap (INDO) schemes. These counter a defect in CNDO which results in there being no distinction between singlet and triplet electronic states. These have respectively two electrons paired in one case and of parallel spin in the other. Their energy difference is related to an exchange integral, K. Integrals of the form (ms|ms) are not now neglected if χ_m and χ_s are centred on the same atom.

In the form of INDO used by Pople and co-workers the philosophy of keeping close to *ab initio* calculations is favoured. The more empirical line is followed by Dewar whose variant is given the acronym MINDO. The most recent parameterizations of the latter, AM1 and PM3, do give very satisfactory results when compared with a range of experimental observables.

2.13 Calculated energy properties

Every physical observable has associated with it a quantum mechanical operator, O. The value of an observable or eigenvalue, ω, can in principle be found by using the eigenvalue equation in which the operator operates on the wave function appropriate to the state and condition of the molecule in which we are interested,

$$O\Psi = \omega\Psi$$

Mean values of ω may be extracted from such equations using the formula

$$\omega = \frac{\int \Psi * O\Psi \, d\tau}{\int \Psi * \Psi \, d\tau}$$

In the neater and powerful notation of Dirac this equation is written in the form

$$\omega = \frac{\langle \Psi | O | \Psi \rangle}{\langle \Psi | \Psi \rangle}$$

In fact, apart from the case of energy, where O is the hamiltonian operator, H, this pure approach is rarely used and calculated properties are generally based on the orbital energies ε_i and the expansion coefficients of the molecular orbitals in terms of atomic orbitals, c_{ik}.

Molecular energies are particularly important as so many other properties may be inferred from a study of the variation of energy with some molecular parameter such as bond length or angle.

A calculation is run for a particular defined geometrical arrangement of the constituent atoms yielding molecular orbitals and their orbital energies. From these we may calculate the electronic energy, E_{el}. To this must be added the nuclear-nuclear repulsion terms

$$\sum Z_\mu Z_\nu / R_{\mu\nu}$$

where Z_μ and Z_ν are the charges on the nuclei μ and ν, and $R_{\mu\nu}$ their separation. The total energy E_{total} is thus a smaller negative number than E_{el}.

The electronic energy, E_{el}, can be expressed in terms of integrals involving the molecular orbitals, ϕ_i.

$$E_{el} = \sum_i \varepsilon_i^N + \sum_{i<j} J_i - \sum_{i<j}' K_j$$

Here

$$\varepsilon_i^N = \int \phi_j^* H \phi_j \, d\tau$$

and represents the energy an electron in orbital ε_i would have were it the only electron in the molecule, i.e., its kinetic plus nuclear attraction terms;

$$J_{ij} = \iint \phi_i^*(1)\phi_i(1)\frac{1}{r_{12}}\phi_j^*(2)\phi_j(2)d\tau_1 d\tau_2$$

$$K_{ij} = \iint \phi_i^*(1)\phi_j(1)\frac{1}{r_{12}}\phi_i^*(2)\phi_j(2)d\tau_1 d\tau_2$$

Exchange terms, K, only occur between electrons of the same spin, indicated by the prime on the summation.

Now orbital energies ε_i are also related to the three terms ε_i^N, J_{ij} and K_{ij}, leading to the convenient result for closed-shell molecules which have no unpaired electrons,

$$E_{el} = \sum_i \left(\varepsilon_i^N + \varepsilon_i \right)$$

This rigorous formulation is used in *ab initio* methods, the ε^N being computed from the orbitals ϕ_i.

For many of the semi-empirical methods the approximation is made that

$$E_{el} = \sum \varepsilon_i$$

Such an assumption may appear gross but it does seem to produce numbers which vary sensibly with other molecular parameters.

Every type of molecular orbital computer program incorporates the calculation of E_{el} and E_{total} within its own framework and this is invariably printed out. It must be remembered however that only the *ab initio* energies have any absolute meaning, being related to a zero of energy where all particles are removed to infinity. Semi-empirical calculations all use ionization potentials as parameters and consequently energies cannot be compared between one type of calculation and another. Within one method and for different arrangements of the nuclei in one molecule the energies may be compared.

Molecular geometry

If we take the simple example of the water molecule, we may run a molecular orbital calculation for any number of relative positions of the atoms H and O. The minimum energy geometry found in all types of calculations is similar to the experimental geometry with an angle between the two OH bonds.

In principle this procedure can be followed for a molecule with any number of constituent atoms but the number of calculations required would soon become prohibitive. In practice the starting values of bond lengths and angles are usually taken from known data from spectroscopy or crystal structures, or from values known for similar species. Small variations are then investigated.

Molecular geometries are more often used to confirm the validity of calculations than calculations are used to determine geometries, since there is an abundance of experimental techniques available for investigating molecular geometry. The principles may, however, be carried over to determine other molecular energy variations which are less amenable to experiment.

One such example is the case of vibration frequencies. The energy of a molecule may be calculated as a function of stretching a bond or bending an angle. If the resulting potential curve is then fitted to a suitable expression (even a simple quadratic expansion will do in many cases) then rough estimates of vibration frequencies may be found. Such quantities are on the other hand more of interest to spectroscopists than to medicinal chemists.

An important aspect of the calculation of accurate geometries for small molecules is the current interest in optimizing programs. Such programs calculate an *ab initio* molecular wave function and then optimize the geometry using the gradient of the energy with respect to the $3N$ nuclear coordinates (N being the number of atoms).

Conformation

The molecular energy property which is of most interest to researchers interested in small biologically active molecules is conformation. Questions about molecular geometry can be answered with the aid of molecular models, but these leave open a whole range of conformations because of free rotation about single bonds.

Experimental techniques such as crystallography and nuclear magnetic resonance spectroscopy cannot give more than a small portion of the whole conformational picture.

Quantum chemical calculations may treat conformation exactly like geometry. A calculation is performed for a series of positions of one part of a molecule with respect to another and the energies for each position compared.

If there is only one bond about which rotation can occur then the results may be presented in the form of a curve of energy against angle (Fig. 2.7).

For biological species it is frequently the case (e.g. acetylcholine, catecholamines, histamine, dipeptides) that two rotational angles are required to specify conformations. In this case it is sensible to present the required three-dimensional diagram (energy, torsion angles τ_1 and τ_2) as contour diagrams as in Fig. 2.8. On these energy maps, minima indicate stable conformational structures with the relative depths being indicative of relative stability or enthalpy. The maps also give the heights and shapes of barriers between conformational isomers from which it is possible to calculate rates of interconversion.

As with geometries, the normal computational practice is to start with the bond lengths and angles found for the molecule in the crystal structure and then to hold everything constant except the rotations which are being investigated. This procedure is not ideal. The logical procedure is to do *ab initio* calculations, optimizing both the basis set and the geometry for each point on the potential surface. Thus, in a calculation on the conformation

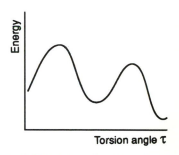

Fig. 2.7 Variation of conformational potential energy with one variable torsion angle.

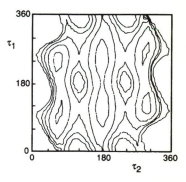

Fig. 2.8 Variation of conformational potential energy with two variable torsion angles presented as equal energy contours.

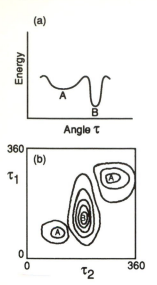

Fig. 2.9 Variations of potential energy with conformation showing hollow minima (A) and deep minimum (B) for (a) one variable angle and (b) two variable angles.

of ethane one would allow for the fact that in eclipsed conformations the H–C–H angles would differ from those in a staggered conformation.

Conformational free energies

The advent of conformational energy maps, largely owing to increasingly rapid molecular orbital calculations, underlines an obvious point of thermodynamics that is often ignored. If there are several minima on a surface then the relative populations depend on free energy differences and not on the calculated energy differences which are internal energies. Entropy effects will frequently be very important. The relative populations of two dips in a potential surface will depend both on how difficult it is to get out of a valley and the ease with which molecules can enter the holes owing to varying widths.

In Fig. 2.9 the case of a wide shallow depression and a narrow deep one are contrasted for both the one-dimensional and two-dimensional conformational diagrams. In both cases the population of molecules with the conformation labelled A with a broad shallow depression of the surface might be more abundant than the internal energy favoured B with a deep narrow hole in the potential surface but with a low probability.

The diagrams also highlight a semantic difficulty in deciding just what area of the map can be referred to as 'conformation' A or B. The boundaries are essentially arbitrary. This point is of relevance to experimental work as well as theoretical calculations since it would remain true even with perfect, real potential surfaces. For this reason many published solution conformation population ratios derived from NMR coupling constant measurements are not very accurate numbers.

If we use some statistical ideas it is possible to compute conformational free energies from the energy maps and hence genuine equilibrium constants. It is reasonable to assume that altering conformation has only a minimal effect on volume so that the maps may be considered as enthalpy, ΔH, surfaces, ΔH being equal to $\Delta E + P\Delta V$. For each region which we define as a conformation, a Boltzmann partition function may be calculated. The partition function Z is a number which indicates how the molecules are spread amongst available energy levels. For the single angle case this would be defined as

$$Z_{\text{conformational}} = \sum \exp\left[-E(\tau)/k_{B}T\right]$$

and in the two variable angle instance

$$Z_{\text{conformational}} = \sum \exp\left[-E(\tau_1, \tau_2)/k_{B}T\right]$$

where $E(\tau)$ or $E(\tau_1, \tau_2)$ is the calculated energy for a conformation defined by τ or (τ_1, τ_2), k_{B} is the Boltzmann constant, and T is the temperature (normally 37°C). The summations have to be taken over a regular grid of points, fine enough to reflect the shape of the surface appropriate for a given conformer. Effectively we are using a numerical form of integration, assuming that the torsional energy levels are very close together. The appropriate portion of the surface can be defined by taking a local minimum point and defining as a conformer all space within a contour set at $2k_BT$ above the minimum. Then

$$\Delta G^\circ = -k_B T \ln \frac{Z^A_{conformational}}{Z^B_{conformational}}$$

There is a further important application of the use of conformational partition functions which is of particular value when we wish to compare the conformational flexibility of a series of similar molecules. We may be interested in knowing for a series of similar, conformationally flexible, compounds, just how flexible they are and what range of conformations is likely at body temperature for each molecule.

This information is contained in the potential energy map but there is so much detail that comparisons are far from obvious. As in computing free energies we can associate with each point on the energy surface a probability

$$Z = \exp\left[-E(\tau_1, \tau_2)/k_B T\right]$$

The probability function can then be integrated numerically over the total surface using Simpson's rule to yield Z and normalized by correcting the points using

$$Z^{new}_{\tau_1, \tau_2} = Z^{old}_{\tau_1, \tau_2}/Z$$

so that the function integrates to unity.

We can now generate probability maps, single-contoured diagrams with the same axes, say τ_1 and τ_2, where the contour will contain within it an indication of a given percentage of molecules for a chosen temperature. If we consider the two-torsion angle case which is appropriate for all the transmitter substances, clearly the square defined by the axes contains 100 per cent of molecules but, as we become more restrictive, we can emphasize just how flexible a molecule is, as in Fig. 2.10.

With these 'population maps' it is possible to see at a glance just how flexible a molecule may be and which regions of conformational space are favoured. Even more usefully we can compare many members of a series by taking, for example, the 99 per cent contour diagram for each compound and ask the question whether, if biological data are available, there is any indication of conformational requirements for activity.

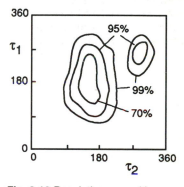

Fig. 2.10 Population map with contours defining the region of conformational space containing defined percentages of molecules for a given temperature.

Ionization potentials and electron affinities

The ionization potential and electron affinity of a molecule should, in principle, be simple to calculate. All that seems necessary is to do a calculation for the neutral molecule in its ground-state geometry and conformation and similar calculations for the positive and negative ions, then by difference we will obtain the desired energies. In practice this obvious path is rarely followed.

One reason why it is not popular is that such a procedure would only be expected to yield values in accord with experiment if the correlation energy error, the relativistic error and the discrepancy from the Hartree-Fock limiting energy were identical for the calculations on the molecule and on the ions. Unfortunately the correlation energy error, being dependent to a first approximation on the number of electron pairs in the molecule or ion will certainly differ in the two situations even though the other two errors may cancel.

A further problem when considering electron affinities but not ionization potentials is that self-consistent field calculations frequently do not converge for negative ions.

In consequence a much more drastic but simply applied approximation is often employed. This equates the ionization potential of an electron with the orbital energy ε_i associated with that electron. This is known as Koopmans' theorem. It implies that not only are relativistic and correlation energies the same in the molecule and in the ion but also that there is no reorganization of electronic structure or distribution on ionization. Clearly such approximations are not valid. Once again, however, when considering differences between ionization potentials for a series of similar molecules many of the errors are constant and an acceptable indication of trends in values on chemical modification can be obtained. The energy of the highest occupied molecular orbital which approximates to the ionization potential is often given the acronym HOMO and used as a parameter in correlations with reactivities.

For electron affinities the energy of the lowest unoccupied molecular orbital, ε_i for the first virtual orbital, may give an indication of the ease with which the molecule may accept an electron. In addition to the assumptions involved in Koopmans' theorem this approximation is made even less acceptable since there is no definite physical meaning for the virtual orbitals. If an *ab initio* computation is run, the virtual orbitals are very variable as a function of the size of the basis set, the only mathematical constraint upon them being one of orthogonality between themselves and with occupied orbitals. The energy of the lowest unoccupied molecular orbital is frequently referred to as LUMO and again is used in statistical correlations.

Electron affinities then are not well calculated although the simple equation of ionization potential with orbital energy is satisfactory for comparative purposes. Absolute calculation requires a lot of work.

2.14 Charge distribution calculations

Possibly more important than nuclear conformation are the details of electronic charge distribution or potential revealed by calculations. Ever since the earliest days of quantum mechanics the square of a wave function at a point in space has been interpreted as a probability. If we have an electronic wave function, Ψ, and integrate this function squared over a volume

$$dv = dx \times dy \times dz$$

then the result will be a sum of the probabilities of finding the electrons in this volume element or an electron density ρ, with

$$\rho = \int \Psi^* \Psi \, d\tau$$

This direct head-on calculation of the charge within a defined volume in a molecule has been used to produce charge distribution data. The problem which hindered the use of the direct approach was the difficulty of doing the integrals for volumes of space which bear an awkward relationship to the coordinate system being used for the molecular calculation.

An attractive way of using the direct integration approach to charge distribution is to plot contours of charge density. This is full of information but has the disadvantage of difficulty of presentation when a molecule has little symmetry.

Mulliken population analysis

A conveniently programmed way of gaining an idea of charge distribution in molecules comes from the so-called Mulliken population analysis. In one line of a computer program we can produce values associated with each atom in a molecule, this figure being the number of electrons 'associated' with the particular atom. In this way all the electrons in the molecule are assigned to nuclei, even though they may not spend much time very close to the particular nucleus. Thus the meaning in physical terms is imprecise although mathematically Mulliken population analyses are clearly defined.

If we are dealing with molecular orbital wave functions,

$$\phi_i = \sum_k c_{ik}\chi_k$$

then the net atomic population of a given atomic orbital, χ_k is defined as

$$P_k = 2\sum_i c_{ik}^2 \chi_k^2$$

Here the subscript i labels the molecular orbital and k refers to the atomic basis function, χ_k. In this we use the concept of the square of a wave function representing a charge density and the atomic basis functions are assumed to be normalized.

$$\int \chi_k^2 \, d\tau = 1$$

The overlap population – the population shared by atomic orbitals k and l – is further defined as

$$O = 2\sum_i c_{ik}c_{il}S_{kl}$$

with

$$S_{kl} = \int \chi_k \chi_l \, d\tau$$

Since $P_k = O_{kl}$ we can write the gross population of atomic orbital χ_k as

$$P_k = \sum_l O_{kl}$$

The total population at any nucleus can be found by adding all the values of P_j for orbitals χ_j which are centred on atom n.

Most molecular orbital programs incorporate the facility of computing Mulliken population analyses. The numbers so generated are then frequently used in applications to biological problems. It is therefore necessary to add a few cautionary words.

To state the obvious, the only charge on an atomic nucleus is its nuclear charge. This will not change with the environment of the nucleus in a molecule. We are more interested in the charge near to a nucleus but 'near to' has to be defined. In population analyses all the charge is associated with nuclei. The charge between two nuclei is divided equally between the two, even if the atoms have very different electronegativities. A more worrying feature of population analyses emerged if *ab initio* molecular orbital functions are used. The resulting populations are not invariant to the basis set. In particular if the basis set is gradually increased in size the results may become bizarre. It is even possible to produce negative populations since S_{ij} integrals can be both positive and negative. Experience has shown that the picture of charge distribution derived from minimal basis set computation is frequently more realistic in terms of accord with experiment than when an extended basis set is employed.

These observations are included to urge caution when using population analyses. They are not useless by any means, but their meaning is not quite as clear-cut as is often inferred from the presentation of a diagrammatic molecule together with a number (the net charge) associated with each atom.

Despite its weaknesses the rough pattern of charge distribution is indicated by a population analysis. Plots of population analyses against rigorously computed charge densities are linear. Above all, when comparing charge distributions of similar molecules the analysis is valuable in indicating trends. Differences are more meaningful than the absolute values.

Molecular potential fields

In some ways more revealing than even an accurate picture of the charge distribution in a molecule would be an indication of the molecular potential field. The molecular electrostatic potential is taken as the interaction energy between a unit positive charge and the unperturbed molecular charge distribution. The latter is due to negative electrons and positive nuclei, so that the electrostatic molecular potential $V(k)$ at a point in space labelled k is

$$V(k) = -\int \rho(1)\frac{1}{r_{1k}}\,dv_1 + \sum_\alpha \frac{Z_\alpha}{R_{\alpha k}}$$

where Z_α is the nuclear charge of nucleus α.

The first-order electron density $\rho(1)$ may be derived from an *ab initio* calculation by taking the occupied molecular orbitals and squaring them, i.e.

$$\rho(1) = \sum_{\text{occ. mo}} n_i \phi_i^*(1)\phi_j(1)$$

A full calculation of this nature can provide what is sometimes referred to as an 'exact' potential. However, more frequently, further approximations are introduced. Often semi-empirical wave functions are used and the electronic distribution replaced by a set of point charges computed from the wave function, perhaps by the use of population analyses.

Because of the speed with which potential field computations may be carried out using gaussian basis functions it becomes possible to treat large molecules of pharmacological interest and to present the results in the form of isopotential maps. The potential maps represent contours connecting points at which the energy of interaction of the unperturbed molecule with a proton is identical or 'isopotential'.

Sensible atomic charges may be obtained by the fitting of the molecular electrostatic potential to that obtained from effective charges on individual atoms.

Frontier electron density

The frontier electron theory was originally developed to explain the difference in reactivity at each position in an aromatic hydrocarbon. It is based on the intuitive idea that the reaction should occur at the position of the largest density of the electrons in the frontier orbitals, which are defined according to the type of reaction:

(a)   In an electrophilic reaction, the highest occupied molecular orbital (HOMO).
(b)   In a nucleophilic reaction, the lowest unoccupied molecular orbital (LUMO).
(c)   In a radical reaction, both of these.

This theory was later given a sound theoretical basis by Fukui, who then introduced the concept of superdelocalizability. Denoting the occupied molecular orbitals by 1, 2, ...m, and the unoccupied levels by m+1, m+2, ...N the superdelocalizability, S_r, is given for the three types of reaction by:

(a)   for an electrophilic reaction

$$S_r^{(E)} = 2 \sum_{j=1}^{m} \frac{c_{rj}^2}{\lambda_j}$$

(b)   for a nucleophilic reaction

$$S_r^{(N)} = 2 \sum_{j=m+1}^{N} \frac{c_{rj}^2}{\lambda_j}$$

(c)   for a radical reaction

$$S_r^{(R)} = \sum_{j=1}^{m} \frac{c_{rj}^2}{\lambda_j} + \sum_{j=m+1}^{N} \frac{c_{rj}^2}{\lambda_j}$$

where c_{rj} is the coefficient of the rth atomic orbital in the jth molecular orbital, and λ_j is the coefficient in the orbital energy, which is given as $\varepsilon_j = \alpha + \lambda_j \beta$ (α is an ionization potential and β an empirical energy parameter used in simple π-electron Hückel theory).

The orbital which mainly determines the value of S_r in each type of reaction is the same as the frontier orbital previously considered.

There are problems in the use of both frontier electron density and superdelocalization. The latter concept was originally put forward considering the π-electron part of the molecule only, with the energies of the orbitals being given in units of the resonance integral of a C–C bond in benzene, β. This means that in a series of molecules there would be a common zero of energy. Using all-valence molecular methods, the energies are obtained in absolute terms, so that the zero of energy is in the unoccupied orbitals. This is obviously not constant in a series of molecules.

The frontier electron density strictly permits only a comparison of reactivities at different positions within the same molecule. In order to extend this concept for use over a series of molecules, a further quantity, F, may be considered:

$$F = f_r / \varepsilon$$

where f_r is the frontier electron density, ε is the energy of the appropriate frontier orbital.

F may be thought of as a weighted frontier electron density, in the sense that ease of removal of the particular electron is also considered.

Frontier orbital theory may be made more rigorous if *ab initio* sphere charges which have been described above are used to provide the charge in specified regions of particular orbitals. Particularly for comparative purposes it is simple using freely available programs to calculate the charge in a sphere of defined radius on an atom, in a bond or indeed at any suitable point in space. This charge may be broken down into orbital contributions by giving an occupation number (the number of electrons in the orbital) equal to zero for all molecular orbitals other than the one of interest. The orbital of interest is likely to be the HOMO if the molecule is thought to be donating charge but LUMO if it is postulated to be acting as an electron acceptor.

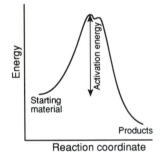

Fig. 2.11 Profile of energy of a chemical reaction. Static indices usually refer to the starting materials not the more relevant transition state.

2.15 Static indices

The use of charges, frontier electron densities and other static indices derived from calculations on an unperturbed molecule as a guide to its reactivity is very dangerous. Reactivity is dependent on transition states and not on the unreacted starting materials as illustrated in Fig. 2.11. All that the calculations on an isolated species may indicate which is of any relevance to the reactivity is the initial slope of the curve.

2.16 Summary

Quantum mechanics provides molecular wave functions in the form of coefficients which multiply known basis functions. Most applications involve computing the energy of a molecule for a given arrangement of atomic nuclei. The calculations of energy properties are now of comparable accuracy to experimental determinations. This is certainly true for geometries and static properties, but reactivities are more of a problem.

The methods have developed to the point where virtually any question about an isolated molecule can be answered by quantum mechanical calculation be it *ab initio* or semi-empirical. In molecules in solution however the techniques of statistical mechanics become essential.

3 Molecular mechanics

3.1 Introduction

The calculations outlined in the previous chapter, while offering a rigorous description of the molecule from both a structural and electronic point of view, suffer as a result of their computer requirements. The amount of computer time to do conformational analysis, including minimization, or to define a potential surface even for a small molecule, can be prohibitive in terms of both time and memory. When these problems are considered in tandem with the explosion of structural data in both the biological and materials fields, it is obvious that a different representation is required. In addition, it should be borne in mind that for many applications the detailed information supplied by the quantum mechanical calculation is of little interest, especially if only a molecular geometry is required. At all times one should be pragmatic as to which method would be the best for a given application.

Just as the 1960s witnessed an explosion in the application of quantum mechanical methods, there was a complementary interest in so-called 'force field' methods for conformational analysis. The origin of these methods lies in vibrational spectroscopy, where the information derived from detailed analyses of vibrational spectra required the development of potential functions to describe the overall molecular behaviour. Two different approaches were considered. In the first, the Central Force Field (CFF) method, the molecular vibrations were fitted to a function which was a sum of pairwise interactions, without reference to the covalent structure of the molecule. The obvious disadvantage of this approach is that although such a description is correct in terms of a quantum mechanical model of a molecule, it lacks the intuitive link with structure with which chemists are more happy. The second method, the Valence Force Field (VFF), provides such a description in that the vibrational data is fitted to a potential function consisting of bond length and bond angle dependent terms. This is much more satisfactory and has the advantage of allowing comparisons between molecules; unlike the CFF potential functions which will be very molecule dependent. The major criticism of the VFF method is that the force constants produced must attempt to incorporate intramolecular interactions such as dispersion forces which result from electron correlation, and therefore are not simply a representation of the intrinsic vibrational frequency.

These spectroscopic force fields provided the ideal starting point for what is now called molecular mechanics. By bringing together features from both the CFF and VFF methods, it proved possible to derive energy functions which were at once chemically intuitive while still retaining the concept of through space attractions and repulsions.

The theoretical basis of the molecular mechanics method can be derived by taking an alternative approach to the Born-Oppenheimer approximation to that considered in molecular orbital methods: in this case the nuclear motion is considered while implying a fixed electron distribution associated with each atom. To this end a model has been developed whereby a molecule is represented as a collection of spheres (possibly deformable) joined by springs. The motions of these atoms can then be described by the laws of classical physics and simple potential energy functions can be used. This allows much larger chemical systems (of the order of thousands of atoms) to be investigated.

Although this method of calculation sounds ideal, the following caveats must be kept in mind. First, as the method neglects explicit representation of electrons, it is restricted principally to the discussion of molecular ground states. This also disallows the investigation of reactions. Secondly, the results obtained will only be as good as the potential functions and parameters used; much of the potential surface defined by the force field has little validity as, typically, only extrema (stable conformations, rotational barriers etc.) are used in the parameterization procedure.

3.2 The energy calculation

The molecular representation introduced in the previous section was one which treated molecules as a set of vibrating spheres. The next step is to define an energy function which is consistent with this concept yet allows accurate calculation of molecular properties. The force fields commonly encountered today have resulted from a number of generations of development. Typically, more structural and thermodynamic data have become available, coupled with considerable increases in computer power, allowing an extension of the functional form of the energy calculation. As more terms are included, the accuracy of the force field increases.

The energy of a molecule is calculated as a sum of the steric and non-bonded interactions present. Therefore each bond length, angle and dihedral is treated individually while non-bonded interactions represent the influence of non-covalent forces.

$$E_{tot} = E_1 + E_\theta + E_\omega + E_{nb} \tag{3.1}$$

Here E_1, E_θ, E_ω, and E_{nb} are respectively the total bond, angle, dihedral and non-bonded energies. This is shown pictorially in Fig. 3.1.

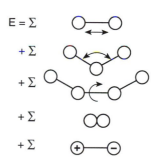

Fig. 3.1 Pictorial representation of the terms included in a molecular mechanics force field.

Bond stretch

The typical vibrational behaviour of a bond is near harmonic close to its equilibrium value but shows dissociation at longer bond lengths (Fig. 3.2). The most accurate description is the Morse function

$$E_1 = \sum D_e[1 - \exp\{-\alpha(l - l_0)\}]^2 \tag{3.2}$$

where l_o is the equilibrium bond length, D_e the dissociation energy, and α a force constant. However, the exponential calculation is computationally expensive therefore most force fields have adopted a simple harmonic function

$$E_1 = \sum k_1 (l - l_o)^2 \tag{3.3}$$

k_1 being the stretching force constant describing the deformation. The bond stretch is treated in the same fashion as a stretched spring. This equation has obvious limitations in that it only approximately describes the actual behaviour of the bond. Further, at extended bond lengths it is much too steep (see Fig. 3.2), while it provides no representation of dissociation at very large deformations. When discussing minimization of poor geometries we will see that this can be an advantage as this function can allow more extended bonds to remain intact.

Other variations on Eqn. 3.3 have been used to accommodate more accurate long distance behaviour. Most commonly this takes the form of an additional cubic term

$$E_1 = \sum k_1 (l - l_o)^2 + k_1' (l - l_o)^3 \tag{3.4}$$

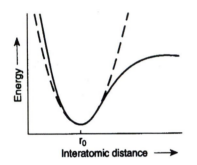

but this suffers from the problem of inversion at long distances. Attempts have been made to remedy this by adding a quartic term which reverses the inversion.

Fig. 3.2 Curves showing the variation of bond stretch energy with distance: — Morse potential; -- harmonic potential.

Bond angles

Historically, bond angles have been treated in the same way as bond lengths and are usually described by a harmonic function.

$$E_\theta = \sum k_\theta (\theta - \theta_o)^2 \tag{3.5}$$

As before, k_θ is a force constant and θ_o the equilibrium value for the bond angle. Again, this term is not ideal for the full range of values observed so higher order terms must be added. In very strained ring systems, however, it is usually not possible to use the constants derived for unstrained and acyclic molecules so separate three- and four-membered ring constants have been developed.

Dihedral angles

In very early force fields it was thought that this term could be omitted; gauche-trans energy differences would then result from non-bonded interactions. This soon proved to be an impossible task and dihedral angle terms were explicitly included. The functional form of this term is a Fourier series

$$E_\omega = \sum V_n (1 + s \cos n\omega) \tag{3.6}$$

where V_n is the rotational barrier height, n the periodicity of rotation (e.g. in ethane n = 3; in ethene n = 2) and $s = 1$ for staggered minima and -1 for eclipsed minima. Fig. 3.3 shows the n = 1, 2 and 3 curves.

In the simple molecules above, a single term, summed over all interactions, would suffice, but as the symmetry across the rotatable bond breaks down, the complexity of the energy profile increases. These can be corrected by the inclusion of other Fourier terms. Consider butane as 1,2-dimethyl ethane. Obviously the eclipsing interaction of the two methyl substituents will be higher in energy than that between one methyl and one hydrogen, or between two hydrogens. Increasing the size of the C–C–C–C three-fold barrier (V_3) would artificially modify the energy of the other eclipsing interactions. Modifying the interaction by the inclusion of a onefold term (since the methyl-methyl eclipse occurs only once per 360° rotation) is the only option available. By way of warning, it should also be noted that the V_n parameters do not represent the complete rotational barrier but that van der Waals interactions must also be taken into account.

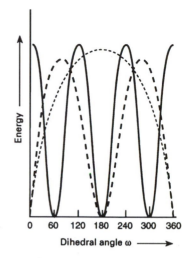

Fig. 3.3 Variation of energy with dihedral angle for one-($\cdots$), two-(--) and threefold (—) barriers.

Non-bonded interactions

The interactions discussed in the previous sections can also be grouped together as the bonded interactions, in the sense that they are defined by the connectivity of the molecule. The non-bonded interactions, on the other hand, are distance-dependent and are calculated as the sum over all atoms with a 1,4 or greater separation. It is usual to consider these interactions as having two components: van der Waals and electrostatic. The former can be considered as both a size parameter and representative of electron correlation (resulting from instantaneous dipole interactions), while the latter provides a quantitative measurement of the influence of polarity on the energy and structure.

Many different functional forms have been used for van der Waals interactions but the most common is the so-called 6–12, or Lennard-Jones potential

$$E_{vdw} = \sum \epsilon[(r_m / r)^{12} - 2(r_m / r)^6] \qquad (3.7)$$

ϵ is the well depth and r_m is the minimum energy interaction distance (Fig. 3.4). Short range repulsions are accounted for by the r^{-12} term whereas London dispersion-attraction forces are mediated by the r^{-6} component. At short distances the repulsive term dominates. The theoretical validity of this function is discussed in most physical chemistry textbooks.

Other forms have been proposed for the van der Waals interaction, principally because the r^{-12} term can be too steep at just less than optimal distances; these short contacts can be important when investigating sterically crowded structures. In the Buckingham potential

$$E_{vdw} = A\exp(-Br) - Cr^{-6} \qquad (3.8)$$

an exponential replaces the repulsive r^{-12} term. In most circumstances this function behaves similarly to the Lennard-Jones equation but at very short interatomic distances the function inverts and goes to $-\infty$, an obvious danger in poorly constructed model structures.

The choice of function has tended to be driven by computing requirements. For a small molecule the number of interactions is relatively small and the close range behaviour is crucial. In this situation the overhead involved in calculating r, as opposed to r^2, and the exponential is not that high. For a protein the number of interactions is considerably higher, but the contacts are likely to be close to equilibrium values. Using the Lennard-Jones function avoids the calculation of large numbers of square roots and exponentials (r^{-6} can be calculated from r^2). The 6–12 function also has the added advantage of requiring fewer parameters.

The second component of the non-bonded potential is the electrostatic term. This is usually calculated using partial charges (q) on the atom centres with the energy calculated using Coulomb's law

$$E_{el} = \sum q_i q_j / D r_{ij} \tag{3.9}$$

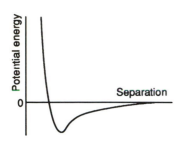

Fig. 3.4 A typical van der Waals curve.

with the dielectric constant D taking a value appropriate to a given solvent or made proportional to the distance r_{ij} between the charges. The electrostatic contribution is one of the most controversial in molecular mechanics and will be discussed further as part of the parameterization procedure.

Other terms

The five terms outlined above constitute the core of almost all molecular mechanics force fields; in some case the entire energy function. In many situations, however, it is necessary for additional terms to be included.

For systems where hydrogen bonding is vital for stability, e.g. biological molecules, it has been common to include an additional, explicit hydrogen bond energy function to ensure correct geometries. In certain protein force fields this takes the form

$$E_{hb} = \sum (C_{ij} / r_{ij}^{12}) - (E_{ij} / r_{ij}^{10}) \tag{3.10}$$

Other force fields attempt to simulate hydrogen bonds using just the van der Waals and electrostatic terms without the inclusion of a special attractive potential. This latter method could well be the more valid as, in an attempt to retain optimum hydrogen bonding geometries, the explicit function might give correct configurations at the expense of creating strain elsewhere.

A second problem which can arise is restricting the planarity of isolated unsaturated centres. The four atoms in this grouping should be kept in a plane, however, the branch atom – e.g., in the case carbonyl groups, the oxygen – can be distorted. If the distortion is measured as the height of the central atom above the plane formed by the other three atoms then a simple restraining force can be used to hold the group in its correct geometry.

$$E_{opl} = \sum k_{\chi} \chi^2 \tag{3.11}$$

where k_χ is the force constant and χ the height above the plane. A modified version of this type of function can also be used as a chirality constraint.

Thus far, all of the potential functions have been concerned with isolated features of molecules. If any structural changes are correlated then, in this type of force field, they must result from a combination of appropriate forces. If only structural and thermodynamic data are required to be reproduced, this form of the force field (the so-called second generation) is adequate. However, to fit, in addition, vibrational frequencies, the coupling between geometric features must be explicitly included in the representation of the molecule.

If one considers the structure of butane (Fig. 3.5) it is clear that as the conformation changes from *anti* to *syn* there is a change in the C–C bond length and an opening of the C–C–C bond angles. The best way to incorporate this kind of feature into the force field is *via* a stretch-bend interaction. This has the effect of restraining distortion of the angle through compensatory bond stretches. A potential term of this type allows greater transferability of the isolated bond angle terms which, if fitted to butane data, would otherwise give poor ethane geometries.

Other commonly used cross terms include bend-bend and torsion-bend. The functional forms are as follows:

stretch-bend
$$E_{l\theta} = \sum\sum k_{l\theta}(l - l_o)(\theta - \theta_o) \tag{3.12}$$

bend-bend
$$E_{\theta\theta'} = \sum\sum k_{\theta\theta'}(\theta - \theta_o)(\theta' - \theta'_o) \tag{3.13}$$

torsion-bend
$$E_{\theta\theta'} = \sum k_{\theta\theta'\omega}(\theta - \theta_o)(\theta' - \theta'_o)\cos\omega \tag{3.14}$$

where the k terms are the force constants; l, l_o, θ, θ_o and ω are as before.

A force field including all of the valence, non-bonded and cross terms can be parameterized to give very close agreement with experiment for a large number of properties. Problems may occur, however, when highly polarizable groups or ions are present. In this situation the static charge distribution assumed in the electrostatic term is no longer realistic or adequate and an additional polarization potential will be required. Also, the van der Waals potential assumed strict pairwise interactions but more complex, many-body terms might also come into play. Attempts to include these effects are still at an experimental stage and no simple representation can be prescribed.

The final situation to be considered is the presence of delocalised π systems. This does not have any bearing on protein and other biological systems as the π systems are isolated from other unsaturated regions, e.g. aromatic amino-acids and DNA bases. This allows the use of large twofold barrier constants which will resist deformation out of the plane without interfering with the conformational mobility of other regions of the molecule. For many small, unsaturated molecules, and systems based upon porphyrins, this is not possible, unless one creates a new set of atom types for every instance. Here, some account must be taken of the degree of partial

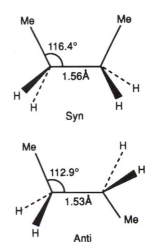

Fig. 3.5 Molecular geometries for *syn* and *anti* butane structures.

double-bond character and this should be allowed to vary according to structural changes. An elegant way of getting around this problem has been implemented in the MM2P program. The molecule in question is divided into conjugated segments and a simple SCF scheme is used to calculate bond orders about this fragment. These bond orders allow new bending, stretching, and torsional parameters to be calculated which are sensitive to the degree of local π character at any given atom. In the simplest case the parameters vary linearly with bond order. A cosine dependence can also be used for the torsion angles indicating the degree of p orbital overlap. As the structure is optimized the parameters for the π system can be recalculated if the geometry changes beyond a certain tolerance.

3.3 Energy minimization

If one has generated a model using molecular graphics, based upon standard molecular fragments, or from a z-matrix of typical internal coordinates, the energy obtained from the molecular mechanics calculation is likely to be high and not representative of the actual structure. To obtain more reliable geometries and energies one must attempt to minimize the energy of the system. This problem can be approached in two ways: either one can vary the actual internal coordinates to find their optimum value, or, as is more common, work in Cartesian coordinate space and optimize the atomic positions subject to the restraining forces generated by the molecular force field. Since most minimization methods require first, and sometimes second, derivatives of the energy the latter method is more convenient and the potential functions are easily differentiated.

A second issue which must be addressed is local versus global minimization. Given that there are three degrees of freedom per atom, for a molecule of N atoms there are 3N-6 variables to be minimized (subtracting those due to rotation and translation). A multi-dimensional problem of this nature is further complicated by the presence of many local energy troughs on the potential surface which are minima in a mathematical sense, however, they are higher in energy than the lowest energy state, or global minimum. This situation manifests itself even in the simple rotational potential for a 1,2-di-substituted ethane (Fig. 3.6). Clearly the gauche ($\omega = 60°$) conformation is stable but the molecule's preference would be for the trans conformation. This almost trivial example highlights perhaps one of the most difficult problems in computational chemistry: how does one find the global minimum (and how can one be certain that it is, in fact, the lowest energy structure).

Many of the minimization programs currently in use today are based upon the mathematical principles of the Newton-Raphson method. This requires first and second derivative information about the energy surface, but a family of algorithms exist which use different approximations to the second derivative matrix (the Hessian).

An important property of the functions used for force field calculations is that they are continuous and differentiable. From simple calculus we know

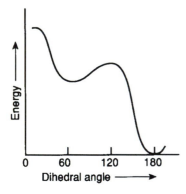

Fig. 3.6 Shape of the general rotational potential for 1,2-di-substituted ethanes.

that the condition for a minimum on a curve, the point x^*, is that the first derivative equals zero, i.e.

$$f'(x^*) = 0 \tag{3.15}$$

Since our starting point is x and not the minimum x^* we can write

$$x^* = x + \delta x \tag{3.16}$$

where δx represents the changes which x must undergo to reach the minimum value. The condition for the minimum can therefore be written in terms of x

$$f'(x + \delta x) = 0 \tag{3.17}$$

and expanded as a Taylor series

$$f'(x + \delta x) = f'(x) + f''(x)\delta x + f'''(x)\delta x^2 + \ldots \tag{3.18}$$

which is also set equal to zero. Truncating the Taylor series after the second order term gives

$$f'(x) + f''(x)\delta x = 0 \tag{3.19}$$

Rearranging Eqn. 3.19 gives the expression for the change which must be made to x to reach the minimum

$$\delta x = -f'(x)/f''(x) \tag{3.20}$$

which can be substituted back into 3.16 to give

$$x^* = x - f'(x)/f''(x) \tag{3.21}$$

This is the simple one dimensional case. When considering molecules, each atom has three degrees of freedom – in x, y and z – and the term $f'(x)$ must be replaced by a 3N x 1 matrix (**F**) containing terms $\delta V/\delta x_i$, the derivatives of the potential energy (V) with respect to a change in coordinate i. The corresponding second derivative matrix is constructed using all the cross-derivative terms ($\delta^2 V/\delta x_i \delta x_j$) involving each coordinate. Since matrices have replaced the single values in Eqn. 3.21, it is no longer possible to carry out the simple division. Fortunately standard computational procedures exist to derive the inverse of the Hessian matrix and the second term in Eqn. 3.21 can be replaced by $\mathbf{H}^{-1}\mathbf{F}$. Here $\mathbf{H}^{-1}$ represents the inverse of the Hessian.

It should be noted that by truncating the Taylor series the assumption is made that the minimum is exactly quadratic in behaviour. For a complex surface this will not hold true far from the minimum but will be a better approximation as it moves closer. This forces the calculation to be carried out in a stepwise, iterative fashion, rather than reaching the minimum first time.

When working with small molecules (<100 atoms) this approach is efficient and will converge after relatively few steps. As the number of atoms is increased, the number of matrix elements in the Hessian goes up rapidly making the calculation much slower. Additional problems with computer memory requirements for the storage of the Hessian may also be a severe limitation. This has necessitated the application of less efficient, but more practical algorithms, usually through an approximation of the Hessian. These include neglecting off-diagonal interactions between atoms (block diagonal Newton-Raphson), and the diagonal Newton-Raphson method which only calculates $\delta^2 V/\delta x^2$ values, neglecting correlation between the three degrees of freedom for a given atom.

A much more severe approximation to the Hessian is to consider it as a constant. This produces the steepest descent method which is driven purely by force gradients along the potential surface. If one imagines the energy surface as being rather like a hilly landscape then the most reliable way to find a valley, or energy well, is to follow the gradient downhill. As the gradient method has no information about the local curvature of the energy surface, minimization by this method slows down considerably as the gradient decreases. Close to the bottom of the potential well the energy differences can be rather small, however, the forces acting on the molecule can still be relatively large compared to those obtained by a less approximate method. The principal advantage of the steepest descent method is that it is very efficient when very large forces are present, far from a minimum, making it a robust choice to tidy up model geometries prior to further refinement by another method.

It is possible to exert more control in first derivative methods by including some kind of history of the minimization path. In pattern search methods the previous step is used to accelerate movement if the gradient is in the same direction as before. If the gradient changes, the pattern is abandoned and a new one set up.

A more elegant improvement to this method is that of conjugate gradients which, like pattern search, uses information from previous steps to modify the move in the next step. Unlike pattern search, it does not abandon the history if the direction changes. In the first step, where the gradient vector is $\mathbf{g}_1$ the move is given by

$$\mathbf{s}_1 = -\mathbf{g}_1 \qquad (3.22)$$

The new direction from this step takes into account the previous gradient and follows the search direction

$$\mathbf{s}_k = -\mathbf{g}_k + \mathbf{b}_k\mathbf{s}_{k-1} \qquad (3.23)$$

where $\mathbf{s}_{k-1}$ is the search direction from the previous step and $\mathbf{b}_k$ is a scaling factor given by

$$\mathbf{b}_k = \mathbf{g}_{i+1} \cdot \mathbf{g}_{i+1}/\mathbf{g} \cdot \mathbf{g}_i$$

The effect of the second term in Eqn. 3.23 is to release the constraint that the $i+1^{\text{th}}$ gradient should be orthogonal to the ith gradient. Hence, better search directions are usually obtained by this method than by steepest descents and it has much better convergence properties (Fig. 3.7).

In addition to its speed and memory requirements, the conjugate gradients method has the additional advantage that it is unlikely to maximize rather than minimize. This is one of the shortcomings of the Newton-Raphson family of methods which use only derivative information to search for stationary points: this could be either a maximum or a minimum. If a maximum on the energy surface is close to the starting point it is likely that it will be found, and a transition state structure will be obtained rather than a stable ground state.

In conclusion, the choice of minimizer should be tempered by the state of the starting structure: if it is a model, far from a minimum, then a method such as steepest descents should be applied. Once a low energy has been obtained it is advisable to switch to a method with better convergence properties – some variant of the Newton-Raphson method or conjugate gradients – which will have greater success in actually finding the minimum. Applying Newton-Raphson methods to poor structures could lead to catastrophic results, including maximized structures.

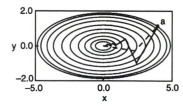

Fig. 3.7 Minimization paths for a simple energy surface: — steepest descent; -- conjugate gradients.

3.4 Force field parameterization

It is obvious that the quality of a given force field depends crucially on the parameters which are derived as constants in the potential function. Just as a given application can dictate which functional form is the most appropriate, the parameterization procedure must include this data when developing force constants and equilibrium values, in order to be valid. This can be both a strength and weakness of the force field method; the predictive ability of the method cannot easily extend beyond certain predetermined bounds, while within these limits the results can be very good indeed.

When discussing the different terms incorporated into the energy function, force fields were referred to as second or third generation. This reflects both the functional form of the potential equation and the scope of the data used in the fitting procedure. Available computer power must be considered as a vital factor in this historical division; obviously some kind of balance must be met between accuracy of calculation and time taken. Another consideration is the types of application of the force field. In the 1970s the principal interest was structure, which can be quite well described by the simple potential forms. More recently, interest has shifted to the dynamic behaviour of molecules. No longer is our interest restricted to the minima on the energy surface and this requires better defined descriptions between the extrema. Only by using complex energy functions which incorporate correlation between internal coordinates has this improvement be achieved.

A parameterization can be approached from two directions. One can attempt to automate the procedure and use least-squares optimization methods to obtain a simultaneous best fit of calculated results to experimental data. Like all multi-dimensional problems, local minimization is a hazard leading to mathematical solutions which have little physical

The development of the MM3 force field, a third generation potential, is discussed in Allinger, N. L., Yuh, Y. H., and Lii, J.-H. (1989). *J. Amer. Chem. Soc.*, **111**, 8551–8566.

significance. Optimization in this way must be closely watched to avoid excessive bias towards some parameters at the expense of others.

Alternatively, a trial and error procedure can be used. Here the user makes small changes to the parameters in an attempt to achieve the best possible fit. One advantage of this constant intervention is that a greater feel is obtained for the inter-relationships between parameters. In combination with a limited least-squares optimization better results should be possible than by the completely automated method.

As stated earlier, the data to be fitted should reflect the kind of applications for which the program will be used. For small molecules the most reliable source of data on geometries is gas phase structural studies, usually microwave or electron diffraction. The use of solution phase or crystal geometries is to be avoided as the influence of the environment is an unknown factor. Spectroscopic data can also provide information on rotational barriers and vibrational frequencies. Thermodynamic data such as heats of formation can also be included, but group contributions to ΔH_f must also be derived. Data from *ab initio* molecular orbital calculations can also provide target values if few experimental data are available. More recently, the latter method has been extended. *Ab initio* molecular energy surfaces are calculated for typical molecules using basis sets of reliable quality. Rather than using equilibrium geometries, these surfaces and their first and second derivatives have been evaluated for a collection of geometries distorted along the normal modes of vibration. These surfaces now describe not just minimum energy positions but give a more accurate representation of the complete potential energy surface. By fitting a potential energy function and its derivatives to this kind of data one can be sure that energies of non-minimum energy structures are considered as 'real' behaviour. The obvious limitation of this approach is that the surface is only as good as the *ab initio* calculations used in its derivation, but with enough computer time this ceases to be a problem. The main problem remaining is that the attractive part of a van der Waals function is not amenable to calculations at the Hartree-Fock limit.

For more details on using *ab initio* surfaces to generate parameters see Maple, J., Dinur, U., and Hagler, A. T. (1989). *Proc. Natl. Acad. Sci., USA*, **85**, 5350–5354.

The next issue is the origin of the starting values for each of the parameters. The important thing which must be remembered is that the resultant geometries come from the total interactions of the force field and not simply the force constant and equilibrium values for a given internal coordinate. This has the serious disadvantage that it is not strictly correct to place physical significance on the components of a force field generated energy. However, this allows greater flexibility in fitting the experimental data as the equilibrium values (l_o, θ_o etc.) become additional variable parameters.

A reasonable starting point for the bond-stretch and angle-bend force constants would be a consensus value from all those that are available in the spectroscopic literature; standard equilibrium values for bond lengths and angles must also be taken. For dihedral angles it should be remembered that there is an additional contribution from the non-bonded energy terms to the rotational potential. This must be subtracted before fitting V_n values.

Finding suitable values for van der Waals parameters presents one of the greatest problems in force field development. Although the close-packed geometry of a crystal would appear to provide details about interatomic distances, a simple deconvolution into atomic contributions is not easy; the close approach of hydrogen in an alkane crystal is mediated not just by H··H interactions but by C··C and C··H pairs. The resultant intermolecular arrangement is a balance over all the interactions at short range. Using heat of sublimation data, however, the crystal energetics can be calculated as a test of the parameters. Typically, however, the van der Waals radii are used as 'soft' parameters, that is to say, the exact values have less influence upon the final geometry than the high force constant bond-stretch and angle-bend terms. Therefore, these parameters can be used to fine-tune the potential.

The electrostatic interaction is unusual in that the charge density will be different for a given atom type across a range of compounds. One must either assign generic values to an atom type and make the assumption that an atom's contribution to the overall electrostatic energy is modulated by the other charges, or one can be consistent as to which method is used to calculate starting charges in the parameterization. Only charges calculated by this method will be compatible with the given force field.

Unless dealing with near ionic systems, the charges on a molecule will have less influence on the final geometry (since the electrostatic force falls off as r^{-2}) than on the energy of a system. This allows one to think in terms of molecular charge distributions rather than absolute values. In some applications (free energy calculations – Chapter 4; molecular similarity – Chapter 6) vast improvements have been found when using better quality charge distributions.

The charges themselves can be calculated by a number of methods. Empirical charge schemes, which have usually been parameterized to give close agreement with experimental dipole moment values, provide a fast way of generating charges. These are certainly suitable for structural calculations and large systems. Some force fields have been parameterized using charges obtained from molecular electrostatic potentials (see Chapter 2) and this procedure should be followed when applying the program to unknown molecules. Although there can be a tendency to use 'the best possible charges' one must always bear in mind how crucial the absolute value will be to the application. There is little point in wasting computer time on expensive *ab initio* calculations when all one wishes to do is tidy up the stereochemistry of a model structure!

A different approach to generating non-bonded parameters is to fit them to data on liquid properties; the Monte Carlo method can be used for this (Chapter 4). This method has been used to derive parameters for a wide range of molecular liquids indicating that it is a reliable procedure.

3.5 Conformational analysis

One of the most useful aspects of molecular mechanics calculations is that they allow rapid determination of molecular energies. This makes them very suitable for problems of conformational analysis. Since the energies calculated are relative, in the sense that they depend entirely on the connectivity of the molecule, they cannot be compared between molecules. However, when investigating conformational space this does not arise; only internal coordinates will be changed, not connectivities.

The simplest way to carry out conformational analysis is to derive the complete energy surface which results from changing a particular rotational variable. Making the approximation that only conformations close to the maxima and minima of the rotational potential will be observed, rotations about a saturated C–C bond can be reduced to six points. For a small molecule with, say, three rotatable bonds this does not present much of a problem as only 6^3 (216) conformations will be generated. If a larger molecule, containing six rotatable bonds, is of interest the number of conformations increases dramatically to 6^6 (46,656); and this is still a very small molecule. Obviously this kind of detailed grid scan must be modified for general use. Also, if more than two variables are used it is difficult to visualise the energy surface being mapped out.

Most of the solutions to the combinatorial problem of conformational search rely on methods to filter out poor conformations at as early a stage as possible. Reducing the number of values per rotatable bond is an obvious answer but immediately it should be recognised that one runs the risk of overlooking potentially interesting structures. In the case of amino acids, each residue to be calculated can be considered a pair of interdependent variables, with only certain minima from the conformational energy map (Ramachandran map) accessible. By reducing the number of pairs to either minima, or minima within a threshold of the global minimum, one can reduce interesting conformational space to a very few points, of the order of 5–10 per residue. Similar restrictions can be applied to any pair of dihedral angles. A conformational search procedure can then proceed by considering combinations of the minima along the given chain.

A second method of increasing the efficiency of a conformational search is to use a tree-search. Here, the search starts out as a complete grid search but as individual conformations are being generated checks are carried out to eliminate unsuitable structures before completion. Typical tests would include checking for adverse non-bonded contacts. Since many large systems contain cyclic regions, ring closure is a particularly good criterion. As a conformation is being generated it should be possible to tell whether it will ever be capable of cyclizing, based upon the maximum length of a unit in the chain (Fig. 3.8).

Usually the conformations generated by any of the above methods will have rigid geometries. Hence, minimization will be required to produce more realistic structures. If only minima are required then one can proceed as in Section 3.3. If, however, one is interested in mapping out a complete surface, and barrier heights and interconversion pathways are of interest, means must be found to hold dihedrals at a particular value while minimizing all other variables. The simplest way to do this is to modify the rotational potential of the fixed dihedral by applying an offset c_2 to the minimum value and using a

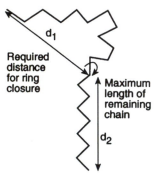

Fig. 3.8 Ring closure criteria for filtering out poor conformations during tree-searches: d_1 must be less than d_2 to continue the search.

very large rotational barrier constant c_1

$$E = c_1[1 + \cos 3(\omega + c_2)] \tag{3.25}$$

Following minimization the energy must be recalculated using the correct rotational potential. Great care must be taken when applying this method, especially when investigating cyclic molecules as the pathway across a surface may depend upon the starting point and the location of other stationary points on the energy surface.

Having obtained a potential energy surface, or conformational reaction co-ordinate, by one of the above methods, it is tempting to place an interpretation on the different contributions to the barrier height etc. This must be approached cautiously as it should be remembered that only features such as total energy, geometry and vibrational frequencies can be considered to have any physical meaning.

In the next chapter we will see that Monte Carlo and molecular dynamics methods, which allow molecules to traverse energy barriers, are more useful for larger systems.

3.6 Summary

Force field methods of the type outlined in this chapter are finding increasing application in many areas of structural chemistry. The principal advantage of these methods is that classically based potential energy functions are less computationally expensive than quantum mechanical methods, allowing systems of the order of thousands of atoms to be considered. A large body of data now exists validating the functional forms and the variable parameters used so that, as long as one is aware of the limitations associated with a given force field, reliable results may be obtained. Many of the older force fields will be valid only for structures while thermodynamic data can be calculated from second generation force fields such as MM2. If one wants to know more about vibrational frequencies, a force field containing cross terms in the energy function must be applied. Typically, this type of force field will be more reliable between stationary points on the energy surface.

Minimization of structures is usually required to remove strain from poorly defined geometries. When far from a minimum, simple gradient techniques, and a less complex force field expression are the methods of choice. But, as one approaches the minimum, more sophisticated methods such as conjugate gradients are recommended: these converge much more quickly. It is also important that the parameters used are consistent with the rest of the force field. Since the parameters arise from a fitting procedure they can have little physical significance, A 'mix and match' approach is to be avoided. Conformational search procedures often utilize molecular mechanics calculations but one must be careful that the methods used do not introduce bias into the calculation otherwise a true representation of the energy surface will not be obtained, or low energy conformers will be omitted.

4 Statistical mechanics

4.1 Introduction

The validity of any computational model results from the successful reproduction and prediction of experimental observables. In the past this has tended more toward gas phase properties and *ab initio* calculations have, in fact, been more accurate in the prediction of spectroscopic frequencies than is sometimes possible by experiment. While serving the purpose of producing reliable structures of isolated molecules, it is clear that many more interesting systems would be amenable to study if calculations could be done in the solution phase. In addition to presenting the problem of how the solvent should be represented, there is the snag that experimentalists will usually produce numbers related to the free energy of a system, whereas the energy calculations considered so far give only potential energies, or at best enthalpies. To overcome this barrier the computational chemist must turn to the techniques of statistical mechanics. In this chapter we shall explore the methods for simulating 'real' systems, and novel methods which can relate the calculated properties of our rather small, model systems to measured bulk properties.

4.2 Solvation

General methods to explore solvation are reviewed in Richards, W. G., King, P. M., and Reynolds, C. A. (1989). *Protein Engineering*, **2**, 319–327.

That solvation plays an enormous role in the determining properties of molecules is self-evident. In physical-organic chemistry the effect of varying the solvent on reaction rates is well known. Also, the conformational equilibria existing in molecules can be changed by varying the medium, which has led to rules based upon favourable orientation of dipoles being dependent upon the dielectric constant of the solvent. In biological systems the major unsolved problem, that of how and why a protein adopts a particular conformation based simply on its sequence, could be reduced to the balance between differential solvation effects about hydrophilic and hydrophobic sidechains. Clearly the environment surrounding a molecule is very important. But how do we represent it?

The simplest way to take into account the influence of the solvent is to make the assumption that its major effect is to screen the electrostatic interactions in the solute. This can be accomplished by including the appropriate dielectric constant in the denominator of the electrostatic term in the molecular mechanics energy function. The calculation is then carried out on the isolated molecule. This method has clear limitations, the most obvious being that it takes no account of favourable solvent-solute interactions. In aqueous solution there could be a number of hydrogen bonds. By neglecting the van der Waals component, contraction of molecular volume can result as the molecule seeks to maximise favourable interactions. In the presence of solvent this tendency is balanced by van der Waals interactions between solvent and solute. It is obvious that, given the short-range nature of the function, its omission will result in distorted geometries. With protein systems the dielectric screening is not always adequate between charged sidechains on the surface. This can produce sidechains lying along the surface of the protein rather than protruding out into solvent.

A potentially much more powerful method is to consider the solute as being a volume of low dielectric constant embedded in an environment with a dielectric constant appropriate to the bulk solvent. The system can now be treated by the methods of classical electrostatics, allowing a more accurate estimation of the electrostatic energy. Systems of this type are governed by the Poisson-Boltzmann equation (Eqn. 4.1) and different, exact solutions exist if the molecule can be assumed to be spherical or cylindrical.

The Poisson-Boltzmann equation is discussed in greater detail in Honig, B., Sharp, K., and Yang, A.-S. (1993). *J. Phys. Chem.*, **97**, 1101–1109.

$$\nabla \cdot \varepsilon(r) \nabla \phi(r) - \varepsilon K^2 \sinh[\phi(r)] + 4\pi q \rho^f(r)/k_B T = 0 \qquad (4.1)$$

Here, $\phi(r)$ is the electrostatic potential, k_B the Boltzmann constant, T the temperature, q the proton charge, ε the dielectric constant and ρ^f the fixed charge density. $K^2 = 8\pi q^2 I/\varepsilon k_B T$; I is the ionic strength.

For a system such as a solvated molecule, the exact geometry of the cavity is not adequately approximated by the analytical solutions to Eqn. 4.1. However, if the atomic charges associated with the molecule are mapped onto a grid, and the grid points are considered to be inside or outside a dielectric boundary, it is possible to obtain numerical solutions to the problem. If the grid separation is close to atomic dimensions (typically 1Å) the results reflect, in some detail, the subtleties of the electrostatic potential. As an iterative solution is obtained there is polarization of the medium by the charges, since the electrostatic potential is calculated beyond the boundary of the molecule. This influences, in turn, the screening effect of the solvent on the solute charges.

Using the electrostatic potential grid the total electrostatic free energy ΔG^a can be calculated by Eqn. 4.2

$$\Delta G^a = \frac{1}{2} \sum q_i \phi^a(r_i) \qquad (4.2)$$

where q_i is charge at point a and ϕ^a is the induced electrostatic potential. Of course, this only gives the electrostatic component of the solvation energy. There is also a component representing the energy expended in creating a cavity in the solvent, and the concomitant reorganization of solvent around this cavity (see Fig. 4.1). The approach currently most favoured is to consider this contribution to be proportional to the solvent-accessible surface area (SA) of the solute molecule using atom-type dependent parameters σ_k

$$\Delta G_{cav} = \sum \sigma_k SA \qquad (4.3)$$

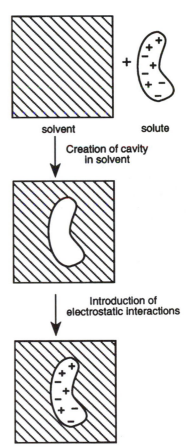

Fig. 4.1 Energy components for the solvation of a charged molecule.

However, the calculation of both terms is necessary only if the absolute solvation energy is required (i.e., starting from the gas phase). When comparing the relative solvation energy changes on going from one solvent to another, it can be assumed that the cavity terms will be approximately equal and, therefore, cancel out.

Similar techniques have also been applied to molecular orbital methods by including continuum modified terms in the Fock matrix (see Chapter 2). In this way the solvation energies and solvent modified electronic properties can be evaluated.

Periodic boundary conditions

Often, a detailed description of solvent behaviour is the motivation for carrying out a simulation. The radial distribution function (rdf) of a solvent shows the probability of finding a neighbouring molecule (m_2) within a certain distance of m_1. By averaging these pair distributions over the complete system i.e., considering each molecule in turn as m_1, a clear picture of the gross solvent structure emerges (Fig. 4.2). If one wishes to observe the response of the rdf to a solute molecule, or the effect of a solute on solvent mobility, then continuum methods are no longer adequate. This kind of study requires the use of explicit representations of solvent molecules in the simulation.

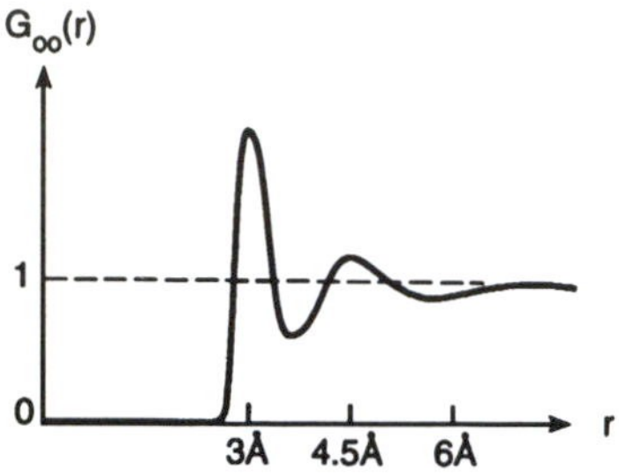

The first obvious disadvantage of this approach is that the size of the calculation is greatly increased. As, for many applications, water is the most interesting solvent, our discussion will be restricted to its representation. Having now increased our calculation to include explicit solvent-solute interactions, it would be expected that all the necessary interactions for reliable simulations will be present. This is not quite the case. The first problem to be overcome is that of system size. All calculations will be carried out on a system of finite size, a box, say. Although the atoms at the centre of the box are receiving their correct complement of interactions those at the faces have a vacuum at one side. This has led to the use of the technique known as periodic boundary conditions (Fig. 4.3). The system of interest is surrounded by images of itself (i.e., the system becomes a type of unit cell in an infinite solution). Molecules close to the edge of the real box can now interact with the image molecules in the surrounding boxes, minimizing edge effects. And, if the motions of a molecule cause it to leave the simulation box an image enters from the opposite face, keeping constant the number of molecules in the system. Care must be taken, however, when choosing the box size and interaction cutoff distances. If the box length is less than twice the cutoff distance, a molecule A could interact with both B and one if its images B', setting up the equivalent of a phonon wave in solid state systems

The second problem concerns the cutoff itself. Using a cutoff of between 8 and 10Å with the various functions representing van der Waals interactions is an adequate approximation given its short-ranged nature. For electrostatic interactions this is not the case. At 8–10Å the electrostatic energy has still to converge, therefore, truncation introduces possible sources of error. Either one must accept the error or use methods which produce shorter range convergence.

Explicit representation of solvent can also suffer from the problem that, if only van der Waals and electrostatic interactions are represented, there is still

$G_{\infty}(r)$

Fig. 4.2 Radial distribution function for liquid water. The first and second solvation shells can clearly be seen at 3Å and 4.5Å respectively.

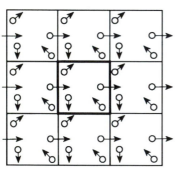

Fig. 4.3 Periodic boundary conditions.

no account taken of the response of the solvent to the presence of charges, i.e., it cannot be polarized. By using more complex water models, including polarization and many-body contributions, more reliable solvation should be accomplished but at the expense of computing requirements.

4.3 Monte Carlo methods

Having set up a system of molecules, we now require methods other than simple energy calculation and minimization to simulate the changes undergone by the system. This can be done using either Monte Carlo methods or molecular dynamics (see Section 4.4).

If we wish to calculate a particular property (Q) of a system with a constant number of particles, temperature and volume (the canonical ensemble) – shortened to (NVT) – classical statistical mechanics shows that the average value of that property is given by Eqn. 4.4

$$<Q> = \int Q(\mathbf{X})P(\mathbf{X}) \, d\mathbf{X} \qquad (4.4)$$

$$P(\mathbf{X}) = \frac{\exp[-U(\mathbf{X})/k_BT]}{\int \exp[-U(\mathbf{X})/k_BT] \, d\mathbf{X}} \qquad (4.5)$$

where $P(\mathbf{X})$ is the Boltzmann weighted probability, $U(\mathbf{X})$ the internal energy, and $\mathbf{X}$ represents all possible states of the system of interest. The problem is how to calculate the probability function $P(\mathbf{X})$ from a collection of molecules acting under a potential field.

The solution to the integrals above could be found by sampling different configurations of the system to provide an indication of all possible states. In its crudest formulation random moves could be made to different molecules. From the energies calculated at each move it would be possible to obtain estimates of $Q(\mathbf{X})P(\mathbf{X})$, which could then be averaged to produce a value of $<Q>$. This approach is clearly flawed since each configuration makes an equal contribution to the configurational integral. Since $P(\mathbf{X})$ is proportional to the Boltzmann factor $\exp[-U(\mathbf{X})/k_BT]$ only configurations with low energies will make a significant contribution to $P(\mathbf{X})$. Hence, many of the configurations generated will have little influence, so proper sampling has not been achieved. The sampling problem was solved in the 1950s by only counting contributions to the configurational integral if they have a probability of occurring proportional to the Boltzmann factor.

If we consider the transition between two states a and b. By first order kinetics the rate of transition p is given by

$$p = \sum -w_{ab}p_a + w_{ba}p_b \qquad (4.6)$$

where w_{ab} is the probability that a transition will take place and p_a the probability that a system is in a given state. At equilibrium $p = 0$ therefore

$$(p_a/p_b) = \exp(-\Delta E_{ab}/k_BT) \qquad (4.7)$$

So the states are related by the Boltzmann constant. Substituting back into Eqn. 4.6 we have

$$(w_{ba}/w_{ab}) = \exp(-\Delta E_{ab}/k_B T) \tag{4.8}$$

Therefore, the probability that a transition will take place is also Boltzmann weighted. We now have a set of conditions which allow correct sampling of all possible configurations of the system: its phase space.

$$w_{ab} \propto \begin{cases} 1, & \Delta E \leq 0 \\ \exp(-\Delta E / k_B T), & \Delta E > 0 \end{cases} \tag{4.9}$$

In practice, when dealing with a solute in a box of discrete water molecules, a new configuration is generated by picking one molecule at random, translating it in three dimensions and applying a rotation about one axis. If the limits for the moves are set too high then the number of acceptable configurations will be too low, so these must be carefully chosen. The energy of the new configuration is evaluated and the criteria in Eqn. 4.9 are applied. If the new configuration has a lower energy than the previous one it is immediately accepted; if it is higher in energy its Boltzmann probability is evaluated and this is compared to a randomly generated number between 0 and 1. If w_{ab} is higher than the random number the new configuration is accepted, otherwise it is rejected and the system is returned to its previous state. By repeated application of this procedure a correctly weighted set of M configurations is obtained, and the configuration integral in Eqn. 4.4 can be changed into a summation giving Eqn. 4.10

$$<Q> = \frac{1}{M} \sum Q(\mathbf{X}') \tag{4.10}$$

where $\mathbf{X}'$ indicates that only configurations with an acceptable Boltzmann weighted probability have been sampled. This is sometimes referred to as Metropolis sampling.

The application of Monte Carlo methods to conformational problems is discussed in Jorgensen, W. L. (1983). *J. Phys. Chem.*, **87**, 5304–5314.

Typically, a simulation would start with an equilibration period to allow proper relaxation of the solvent in response to the insertion of solute. These configurations would be discarded and the simulation would enter a production phase of 500,000–1,000,000 configurations. Here we have assumed that the individual molecules remain rigid throughout. This can be modified to include, say, solute torsional degrees of freedom but the length of the simulation must also be greatly increased to allow solvent relaxation after each change. As molecules are moved at random there is no provision for collective molecular motion.

Monte Carlo calculations are particularly useful when investigating the properties of small molecules in solution as the system is not constrained by the barriers on the potential surface. The random nature of the moves allows the system to move easily from one potential well to another without the requisite kinetic energy to overcome the barrier. However, for larger molecular systems, where there may be cooperative motions about internal coordinates, this method is much less successful. For a molecule the size of a protein very many torsional moves would be disallowed unless they were limited to ridiculously low values. In the latter circumstance it would not be feasible to carry out representative sampling of phase space, hence calculated properties would not be close to convergence. The second disadvantage associated with Monte Carlo simulations is that time-dependent properties such as diffusion coefficients, and rotational and translational correlation functions cannot be monitored. This results from the random nature of the method; the size of an individual move is not controlled by the previous state of the system.

In addition to calculating statistical averages, the Monte Carlo method has found application in searching procedures. In the previous chapter we viewed conformational analysis from a systematic point of view. If, however, random moves are made to the rotatable bonds of a isolated molecule, using the Metropolis sampling condition it should be possible to generate a large number of suitable conformations. The trajectory of acceptable conformations can then be energy minimized, and ranked by energy. The assumption here is that in a user-defined time it is possible to take a representative sample from low-energy phase space. But to reach this convergence point it is still necessary to generate a large number of conformations.

Thus, when interested in structural problems, or in systems with few internal degrees of freedom, the Monte Carlo method is the technique of choice. Otherwise, molecular dynamics may be more suitable.

In the preceding discussion only calculations in the canonical (NVT) ensemble were considered. It is possible to extend the Monte Carlo method to other ensembles by making the appropriate changes to the system. To carry out simulations in the isothermal-isobaric (NPT) ensemble the relationship describing average properties (Eqn. 4.4) becomes

$$<Q> = \int Q(\mathbf{X}, \mathbf{V}) P(\mathbf{X}, \mathbf{V}) \, d\mathbf{X} d\mathbf{V} \tag{4.11}$$

where

$$P(\mathbf{X}, \mathbf{V}) = \frac{\exp[-H(\mathbf{X}, \mathbf{V}) / k_B T]}{\int \exp[-H(\mathbf{X}, \mathbf{V}) / k_B T] \, d\mathbf{X} d\mathbf{V}} \tag{4.12}$$

and the enthalpy

$$H(\mathbf{X}, \mathbf{V}) = U(\mathbf{X}) + PV(\mathbf{X}) \tag{4.13}$$

Sampling is now over all allowed configurations and all allowed volumes. Therefore, the volume of the system must also be subject to a periodic, random change by appropriate rescaling of all the atomic coordinates. As we will see later, this is not as easy to do in molecular dynamics simulations.

4.4 Molecular dynamics

In the chemical sciences molecular structures are usually presented as being rigid except for rotatable bonds. This creates a conceptual barrier when one moves from isolated molecules to 'real' systems as no account has been taken of the fact that the molecule is not in fact rigid, at a minimum energy structure, but is of higher energy, is constantly vibrating along bonds and about bond angles, and is actually somewhere above the fully relaxed potential energy surface. Only when one starts to understand the significance of the dynamic picture of molecules can more complex processes be understood. In biology very large structural changes can take place which would appear to require a massive input of energy. If this process is broken down into a number of smaller changes it becomes apparent that what we are seeing is the cumulative effect of many thermally acceptable excitations. No longer is it necessary, or adequate, to explain these changes as the result of motion of rigid bodies. Unfortunately, many of the changes are not easily observed by experiment, leaving the way open for theoretical methods such as molecular dynamics.

See Karplus, M. and Petsko, G. A. (1990). *Nature*, **347**, 631–639, for a review of the application of molecular dynamics to biological problems.

The equations of motion

A simple definition of molecular dynamics is that it simulates the motions of a system of particles (atoms) with respect to the forces which are present. For an isolated molecule this would mean the valence forces. If we consider a system consisting of hard spheres with position $\mathbf{r}$ acting under some kind of inter-particle potential $V(\mathbf{r})$, say, the Lennard-Jones potential in Chapter 3, then it is possible to calculate the forces which are acting on each member of the system as $-\delta V(\mathbf{r})/\delta \mathbf{r}$. The collection of forces should cause the system to change, but by collective motion of particles over time, in a way that is described by integrating Newton's second law of motion

$$F_i = m_i a_i \tag{4.14}$$

where F is the force acting on a particle, m its mass, and a its acceleration. If we can calculate what the next configuration of the particles is we now have a method to follow the evolution of the system over time. This is distinct from the Monte Carlo method which requires outside intervention in the form of a random move in the system to produce change. In molecular dynamics all changes result from within the system itself, without external intervention.

If we rearrange Eqn. 4.14, and write acceleration as the second derivative of displacement (s) with respect to time, $\delta^2 s/\delta t^2$, we have a more usable version of Newton's law

$$\delta^2 s_i/\delta t^2 = F_i/m_i \tag{4.15}$$

Thus to obtain the dynamic behaviour of our system we must solve this second order differential equation for every particle in the system. Integrating with respect to time gives

$$\delta s_i/\delta t = (F_i/m_i)t + c_1 \tag{4.16}$$

At time $t = 0$ the first term vanishes and the velocity is given by the constant c_1, the initial velocity u_i. At time t we have

$$\delta s_i/\delta t = a_i t + u_i \tag{4.17}$$

in other words the expression for velocity at any time. Further integration with respect to time produces

$$s_i = u_i t + \frac{1}{2} a_i t^2 + c_2 \tag{4.18}$$

where the constant is the current position. This allows one to calculate the displacement from an initial velocity u_i and the acceleration which can be derived from $a_i = F_i/m_i$.

This simple derivation produces an expression which corresponds to the truncated Taylor series for displacement

$$x(t+\Delta t) = x(t) + (\delta x/\delta t)\Delta t + (\delta^2 x/\delta t^2)\Delta t^2/2 + \dots \tag{4.19}$$

therefore a small, persistent error is introduced into the calculation at every time step through the neglect of the higher order terms. Note also that the assumption is being made that the acceleration remains constant throughout the timestep Δt. Unless infinitesimal steps are taken, this is another error-inducing assumption. In practice the time steps used are of the order of 0.5–1 femtoseconds (1 fs = 10^{-15} s), with the restriction that the time difference must be smaller than that for the highest frequency vibration in the system (typically bond stretches). Using a smaller time step with Eqn. 4.19 would produce fewer errors but would require a complementary increase in computer time to allow the simulation of interesting phenomena.

A number of algorithms have been developed to overcome the problems associated with finite time steps and truncation errors. Here we shall discuss the derivation of the Leapfrog Verlet method. If we define v as the average velocity over a time step Δt then the new position is given by

$$x(t+\Delta t) = x(t) + v\Delta t \tag{4.20}$$

Assuming that v is almost equal to the velocity at the midpoint of the time interval

$$v = \delta x/\delta t(t + \Delta t/2) \tag{4.21}$$

we can now calculate v from the midpoint of the previous interval and the average acceleration between $t - \Delta t$ and $t + \Delta t$, so

$$v(t + \Delta t/2) = v(t - \Delta t/2) + a\Delta t \tag{4.22}$$

where a can be calculated from $m^{-1}F(x,t)$. Substituting back into Eqn. 4.20 the new position is given by

$$x(t + \Delta t) = x(t) + v(t - \Delta t/2) + m^{-1}F(x,t)\Delta t \tag{4.23}$$

By avoiding calculating the velocity and acceleration at the same point in time the errors introduced by the truncation can be minimized.

Simulation protocols

The next question is how can we use Eqn. 4.23 to simulate the dynamics of an actual system. Before starting a dynamics simulation the complete system should be minimized to eliminate as many of the poor contacts as possible. Since the algorithm is force driven, the presence of local strain could cause severe problems. The system to be simulated will typically be not unlike those discussed for Monte Carlo simulations. A solute molecule, either a small molecule or a macromolecule, is placed in a box of solvent molecules; edge effects are reduced by the use of periodic boundary conditions. To save computer time, at the expense of accuracy, one can carry out the simulation in the gas phase with an appropriate solvent dielectric constant, or in a sphere of water or in an isolated box. If periodic boundary conditions are not being used it is advisable to divide the system into an inner and outer layer with restraining forces applied to the outer layer of molecules to prevent their 'evaporation'. Having minimized the system, the first stage is to heat it to the required temperature. The temperature of the simulation is calculated from the kinetic energy of all of the atoms in the system

$$\tfrac{1}{2} \sum m_i v_i^2 = \tfrac{3}{2} N k_B T \tag{4.24}$$

but at time $t = 0$ no velocities are known. The system is heated by assigning velocities randomly to the atoms according to a Maxwellian distribution for the given temperature. If a temperature of 300 K is required, a number of intermediate temperatures will be assigned before the system is properly heated. Once the first velocities have been assigned the molecular dynamics method is self-perpetuating; the acceleration is calculated from the forces according to the potential used and the new positions obtained from Eqn. 4.23. In the next step the velocities can be calculated by Eqn. 4.22 and so it continues. During the heating period it is important that the velocities for a new temperature be assigned at random to minimize the development of local, high-temperature 'hot-spots' in the system.

Once the system has the required velocities it is advisable to go through a period of equilibration. This should allow redistribution of the system's energy to ensure stability. Periodic rescaling of the velocities will also be required to bring the system back to the required temperature. The length of the equilibration period should be dependent on the properties being observed: kinetic energy requires a relatively short relaxation period, of the order of a picosecond, while bulk water requires around 10–20 ps. By monitoring system properties such as temperature and the different energy components it should be obvious when they have converged to equilibrium values. Certain programs will not include an explicit equilibration phase but here the user must simply discard the period of the trajectory up until such a time as the system appears stable.

In the final phase the production dynamics is carried out. The exact nature of this phase is determined by which ensemble is being simulated. In the microcanonical ensemble (NVE) the system undergoes free dynamics with no rescaling of velocities. If the system is sufficiently stabilized by the equilibration phase then this is easy to do. However, if the temperature starts to drift there is no way of controlling the system. Drift of this type often occurs as a result of the truncation of the long-range electrostatic forces. If the canonical ensemble is required, close monitoring of the system must be applied. This can be done by adding temperature as an additional degree of freedom to the system. This requires the specification of a 'mass' associated with temperature so there is dynamic transfer of heat to and from the system as required. However, this can set up energy oscillations. A different method is to rescale the temperature at every step using the multiplier λ, where

$$\lambda = \left[1 + \frac{\Delta t}{\tau_T}\left(\frac{T_o}{T} - 1\right)\right]^{\frac{1}{2}} \tag{4.25}$$

Here T_o is the desired temperature and τ_T is a coupling parameter characteristic of the relaxation time of the system. In practice, τ_T is an adjustable parameter which determines how tight the velocity rescaling is; if it is sufficiently weak, the system is not disturbed too much. One disadvantage of this method is that the system no longer exactly corresponds to one of the classical ensembles. Similar procedures can be used to control the pressure of a system, but here it is the coordinates of the atoms which are rescaled.

For a large system (thousands of atoms) a dynamics simulation is very computationally intensive and one is limited as to which properties can be observed. Unless a very fast computer is being used, most simulations will be restricted to hundreds of picoseconds, but even here the time taken will be weeks or months. It was stated earlier that a major limitation on the length of a time step is the high frequency vibrations of bonds. If these could be constrained in some way then it would be possible to use time steps of around 2 ps, allowing twice the length of simulation. Clearly one cannot simply place a high force constant in the bond stretching potential function as this would bias the moves due to other terms. The SHAKE procedure is one of the most commonly used methods to get around this problem. The first stage is to move the atoms as calculated by the Verlet algorithm without considering constraints. It is then possible to calculate corrections to the position of each atom in turn such that they satisfy the condition that the bond length is within some set tolerance of its previous value. Since each bond is considered on its own all the conditions will not be met in one cycle of constraints so a solution must be obtained by iterating over all of the relevant bonds. Studies have shown that these constraints do not alter significantly the calculated trajectory. A typical overhead for the calculation is about 10% which compares favourably with the advantage of obtaining a double length simulation.

In many applications one is interested in only a fraction of the system being simulated e.g., in enzyme-inhibitor studies only the active site and its immediate surroundings are important. Therefore, much of the computer time used is essentially wasted. Fortunately, methods have been developed to overcome this problem and simulations can be carried out using stochastic dynamics methods. First, the system must be divided up into a number of regions (Fig. 4.4). The first region is spherical and centred on the area of

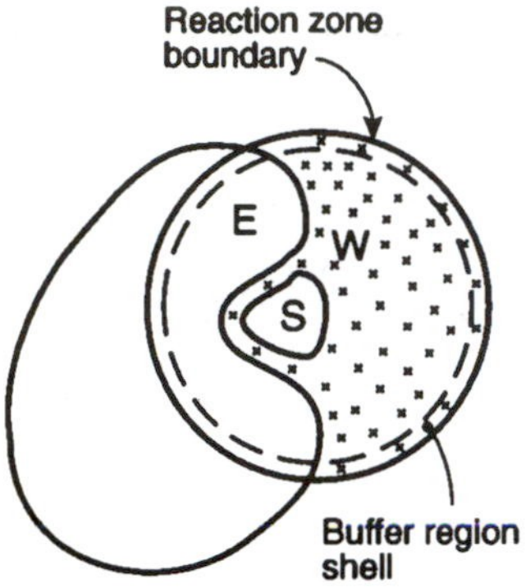

Fig. 4.4 Partitioning of a system for stochastic dynamics. Only those atoms within the circle will be simulated explicitly.

interest; this segment will be treated explicitly in the simulation. In the second region, a shell surrounding the first, the atoms' motions are simulated using a modified equation of motion, the Langevin equation

$$d^2r_i(t)/dt^2 = m_i^{-1}F + m_i^{-1}R - \gamma_i dr_i(t)/dt \tag{4.26}$$

where the first term on the right hand side is as before; the second term allows for the introduction of energy to the main region using the stochastic (random) force R, while the third, frictional term allows dissipation of kinetic energy from the reaction region to the surroundings; in effect representing vibrational damping of the inner regions by the outer zone. This last region of the system is kept fixed in space. In this way the middle region acts as a buffer between the explicitly dynamic portion of the system and the fixed region. This method could cause difficulties, however, when calculating thermodynamic properties; the treatment of long range electrostatic forces is inadequate and temperature and pressure control difficult.

Analysing the trajectory

Having carried out a dynamics simulation, and collected representative structures throughout the trajectory, what information can be gleaned by analysis of the data? The most obvious data are structural: root-mean-squared differences between the starting and final structure, or an average structure over the whole trajectory. If more detailed conformational information is required then plots of individual dihedral angles versus time can show when such changes take place; one of the first applications of molecular dynamics to biological problems was to observe aromatic ring flips in the interior of the protein bovine pancreatic trypsin inhibitor. Fluctuations about average positions can also be correlated with temperature factors from x-ray structures.

One major advantage over Monte Carlo simulations is that time-dependent properties can now be calculated. These can include properties such as diffusion coefficients, which are calculated from mean-square displacements of a particle over time, and correlation functions which give a quantitative measure of the relaxation of features such as reorientation of solvent molecules. Correlation functions have the general form

$$C_A(t) = \frac{\langle A_i(0)A_i(t) \rangle}{\langle A_i^2 \rangle} \tag{4.27}$$

and show the relationship between a property A of a particle at time $= 0$ and time $= t$. This is averaged over all relevant molecules in the system and plotted for a series of times. By fitting this curve to an appropriate function the rate of decay is obtained, and hence the relaxation time. A typical set of water reorientation functions is shown in Fig. 4.5.

Finally, detailed energetic data are immediately available from a molecular dynamics simulation. From a plot of potential energy versus time it is easy to see when a major change has taken place; these should show up as changes in energy. By comparing energy traces to, say, dihedral angle terms from a simulation, we can get some idea of the real barriers for conformational changes. The behaviour of a real molecule is not well described by the minimum energy surface described in Chapter 3 so the barrier heights will not be accurate. Energies extracted from dynamics simulations should show the actual barriers (within the limits of the force field) and, in many cases, a number of barrier crossings will take place allowing estimates in both directions. Molecular dynamics also can be used to generate ensemble averages required for the thermodynamic relationships presented in Eqns. 4.4 and 4.5, if one assumes that, over the duration of a simulation, a correctly sampled distribution of configurations and conformations has been sampled. In principle, this should be the case, but molecular dynamics simulations require energy to traverse energy barriers so the system could remain in a potential well, whereas in Monte Carlo simulations the random changes allow the system to move easily between potential wells. A more detailed discussion of the calculation of free energies from simulation is given in the Section 4.5.

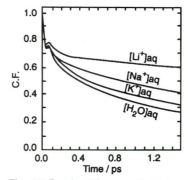

Fig. 4.5 Reorientation correlation functions for water molecules solvating various cations. Fitting an exponential function to each curve allows the differential rotational rates to be estimated.

Structure refinement

The preceding discussion has concentrated on following properties over the duration of the simulation. Molecular dynamics has also proved to be a particularly useful tool in structure determination, most notably NMR. The existence of much higher field spectrometers and the development of sophisticated multi-dimensional NMR techniques now allows the assignment of most protons on a protein of up to 150 residues. Using different experiments one can get an idea of which protons are close in space through the measurement of nuclear Overhauser effects (nOes); the nOe falls off according to r^{-6} so upper limits can be placed on the distances and the signals can be categorized as strong or weak. This distance information can now be used as a penalty function in the potential energy expression such that deviations from the preset limits result in an increase in energy. A simple harmonic function, relating the nOe estimated distance (R_{NOE}) to the distance in the simulation (R_i), is typically used

$$E_{pen} = \sum A(R_i - R_{NOE})^2 \qquad (4.28)$$

with A being the equivalent of a force constant. Carrying out molecular dynamics simulations from arbitrary, model conformations produces an ensemble of structures which satisfy the constraints; most notably, there is greater divergence in the flexible regions of the protein than in the regular secondary structure regions. A similar technique is also used in x-ray crystallography with the penalty function acting on the difference between observed structure factors and those calculated from the simulated structure.

There is one major difference between the dynamics techniques used for structure determination and those discussed earlier. In the former a technique known as simulated annealing is applied; the procedure is similar in principle to that of heating up a metal to very elevated temperatures, where greater motion is possible, allowing more optimum arrangements to be adopted. This is followed by a rapid quench to 0 K. Practically, the model structures are heated up to temperatures of the order of 10,000 K where the structure can traverse a greater range of the available coordinate space than at lower temperatures. This has the effect of allowing the constraint terms to drive the structure to a suitable conformation. The protein can then be cooled by velocity scaling and energy minimized to relieve any residual strain. The advantage of the elevated temperatures is that a wider radius of convergence is obtained for the structures whereas local minima in the potential energy surface would dominate at lower temperatures.

4.5 Free energy calculations

In theory, it should now be possible to calculate the free energy of any system which is amenable to simulation by molecular dynamics or Monte Carlo methods. In practice this is not the case. For a system of particles (or molecules) in the canonical ensemble the expression for the Helmholtz free energy is given in terms of the partition function Z by

A more detailed discussion of the methods in this section can be found in Reynolds, C. A., King, P. M., and Richards, W. G. (1992). *Mol. Phys.*, **76**, 251–275, and Beveridge, D. L. and DiCapua, F. M. (1989). *Ann. Rev. Biophys. Biophys. Chem.*, **18**, 431–492.

$$A = -k_B T \ln Z \tag{4.29}$$

where k_B and T have their usual values. The partition function Z can be written as an integral involving $E(\mathbf{X})$, the configurational energy, over the phase space of the system $d\mathbf{X}$. Thus

$$Z = (8\pi^2 V)^{-N} \int \exp[-E(\mathbf{X})/k_B T] \, d\mathbf{X} \tag{4.30}$$

Therefore, if we could evaluate Z directly, the free energy of the system would be amenable. In fact, it is much easier to calculate relative free energies between processes than to evaluate absolute free energies. This is a result of the problems associated with convergence of the ensemble averages using simulation techniques. First we will show why this is the case. If we generate an expression equal to 1 of the form

$$1 = (8\pi^2 V)^{-N} \int \exp[+E(\mathbf{X})/k_B T]\exp[-E(\mathbf{X})/k_B T] \, d\mathbf{X} \tag{4.31}$$

and insert this into Eqn. 4.29, with inversion of the logarithm we get

$$A = k_B T \ln \left[\frac{\int \exp[+E(\mathbf{X})/k_B T]\exp[-E(\mathbf{X})/k_B T] \, d\mathbf{X}}{\int \exp[-E(\mathbf{X})/k_B T] \, d\mathbf{X}} \right] \tag{4.32}$$

or

$$A = k_B T \ln \left[\int \exp[+E(\mathbf{X})/k_B T] \, P(\mathbf{X}) \, d\mathbf{X} \right] \tag{4.33}$$

where $P(\mathbf{X})$ is the probability function for the system. This can be written more compactly as

$$A = k_B T \ln \, <\exp[E(\mathbf{X})/k_B T]> \tag{4.34}$$

where <...> represents an ensemble average.

This expression can be expanded first, through the exponential, and then the logarithm, to give

$$A = <E(\mathbf{X})> - <E(\mathbf{X})>^2/2!k_B T + \ldots \tag{4.35}$$

highlighting the problem that the absolute free energy is given by an ensemble average of the energy, which converges relatively quickly, but also the energy squared and additional higher order terms. Obtaining convergence of the squared, cubic and higher order terms is not rapid because the higher energy configurations of the system play a greater role. It is precisely these regions of phase space which molecular dynamics and Monte Carlo are designed to neglect.

If we now consider the expression for a free energy difference

$$\Delta A = A_1 - A_0 = -k_B T \ln\left(\frac{Z_1}{Z_0}\right) \tag{4.36}$$

where Z_1 and Z_0 are the appropriate partition functions for the two states, it is possible to derive an exact expression for ΔA.

If we insert a unity in the form of

$$1 = \exp[+E_0(\mathbf{X})/k_B T]\exp[-E_0(\mathbf{X})/k_B T] \, d\mathbf{X} \tag{4.37}$$

into Eqn. 4.36 we get

$$\Delta A = -k_B T \ln\left[\frac{\int \exp[-E_1(\mathbf{X})/k_B T]\exp[+E_0(\mathbf{X})/k_B T]\exp[-E_0(\mathbf{X})/k_B T]d\mathbf{X}}{\int \exp[-E_0(\mathbf{X})/k_B T]d\mathbf{X}}\right] \tag{4.38}$$

therefore

$$\Delta A = -k_B T \ln\left\langle \exp[-\Delta E(\mathbf{X})/k_B T]\right\rangle_0 \tag{4.39}$$

Now the free energy difference can be obtained as an ensemble average over representative configurations of the initial state of the system.

The main requirement for the use of this method is that the energy difference between the two states, $\Delta E(\mathbf{X})$, must be $< k_B T$ for proper convergence. Since many chemical applications will give greater energy differences one must divide the transformation into a number of windows with the total free energy difference arising from a sum over all windows. To do this we require a coupling parameter λ which links the energy expression for the two limiting states. A mixed energy function can be formed to represent intermediate states

$$E_\lambda(\mathbf{X}) = \lambda E_A + (1 - \lambda)E_B \tag{4.40}$$

The coupling parameter is varied from 0 at the start of simulation 1 through to 1 over a number of intermediate simulations, each with a different mixed energy potential. Fig. 4.6 illustrates the windowing procedure.

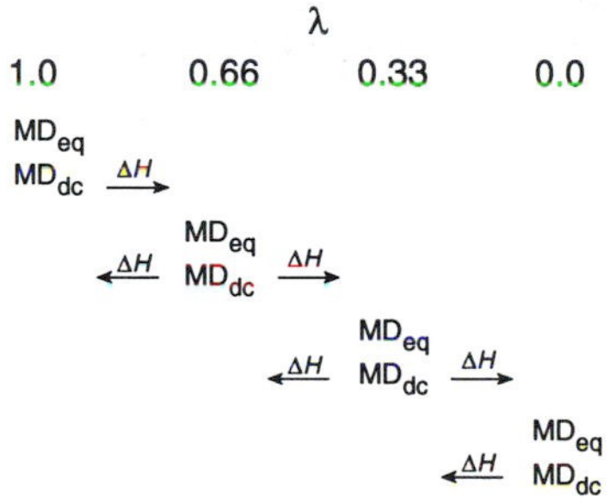

Fig. 4.6 The simulation protocol for 'windowing'. The initial and final states are linked via the two intermediate states defined by λ. For each value of λ the system must be equilibrated (MD_{eq}) prior to data collection (MD_{dc}). For a system at equilibrium the free energy difference can be calculated in both directions simultaneously.

The principal advantage of the perturbation formula (Eqn. 4.39) when used with computer simulations is that there is no requirement that the free energy change must take place over a physical pathway. Figure 4.7 shows the thermodynamic cycle for association of two different molecules S_1 and S_2 with a macromolecule (an enzyme, for example) X. Experimentalists could obtain a free energy difference for the binding of S_1 or S_2 to X. Since free energy is a state function, its value does not depend on the pathway taken, so for the closed cycle in Fig. 4.7 we have

$$\Delta G_1 - \Delta G_3 = \Delta G_2 - \Delta G_4 \tag{4.41}$$

To simulate the binding steps, ΔG_1 and ΔG_3, would be an enormous task as, not just ligand binding, but desolvation of the ligand and the binding site would be required. Using Eqn. 4.39 it should be possible to calculate the non-physical processes on the vertical axes of Fig. 4.7 by using the windowing procedure to effect a transformation of S_1 to S_2 over a number of stages. Thus, the two simulations would be the transformation of S_1 into S_2 in solution, and converting S_1 into S_2 at the binding site. Clearly, as long as S_1 and S_2 are not too different, these changes should not disturb either system duly.

Since its first application to biological systems in the 1980s a vast number of interesting applications have been investigated using this form of computer alchemy. If the transformation of A to B is calculated in both a hydrophobic and hydrophilic solvent, partition coefficients can be obtained. Mutating one residue type to another in a protein allows the estimation of the effect of the mutant on binding or stability. If a substrate is modified in solution and at the active site, binding energies will result, and by mutating the protonated and unprotonated forms of a molecule we can obtain values of pK_a. Many other applications can be found in the Further Reading.

The principal problem with this method is that the two states A and B must differ in such a way that they both affect their environment in similar ways. If the changes are too great then the simulations will have to be longer to allow convergence, or the average over the initial state will not be valid. Depending on the exact nature of the system, either molecular dynamics or Monte Carlo simulations can be used. Obviously each simulation must be long enough to ensure correct sampling of phase space.

A second method for evaluating free energy differences is thermodynamic integration. If we again describe the transformation by a coupling parameter λ, the free energy change can be calculated as

$$\Delta A = \int_0^1 (\delta A(\lambda) / \delta \lambda)\, d\lambda \tag{4.42}$$

From this expression it is possible to derive the relationship

$$\Delta A(\lambda_A \rightarrow \lambda_B) = \sum_N \Delta E \tag{4.43}$$

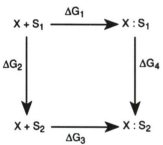

Fig. 4.7 A typical closed thermodynamic cycle for the binding of two different binding agents, S_1 and S_2, to a macromolecule, X.

so that by gradually transforming one state to another over N steps we can calculate the free energy difference as being a potential energy difference. Similar criteria for the magnitude apply as for the perturbation formula.

If the transformation is conformational, it might be more convenient to calculate the free energy difference as a potential of mean force over the conformational degree of freedom. This method is not discussed here but can be found elsewhere.

4.6 Summary

Many of the chemical processes of interest to computational chemists occur not in the gas phase but in solution. This requires an adequate representation of the solvent environment. The simplest methods for the simulation of the solvent are concerned with electrostatic screening and are incorporated in the calculation *via* the dielectric constant. Unfortunately this neglects effects due to solvent packing around the solute and, since the molecule is still isolated, provides no information about the molecular behaviour of the solvent. If the latter is of interest it is necessary to incorporate an appropriate number of explicitly defined solvent molecules into the calculation. The intermolecular interactions can be described by suitable molecular mechanics parameters. Restrictions in computer time will usually limit the number of solvent molecules giving rise to spurious effects at the edges of the finite system. This can be overcome by using a square or rectangular system which can then be surrounded by images of itself such that edge molecules interact with images rather than a surrounding vacuum. This is known as applying periodic boundary conditions.

Two principal methods are applied to the simulation of collections of molecules. Each can be used to generate representative configurations for the calculation of ensemble averaged properties. In the first, the Monte Carlo technique, random moves are made to the molecules and the energy differences tested according to their Boltzmann weighted probability of existing. Only likely configurations contribute to the ensemble average. In the molecular dynamics method all changes are made subject to the inter- and intramolecular forces present in the system, *via* classical equations of motion. This allows time-dependent behaviour such as diffusion coefficients and relaxation times to be monitored.

Although absolute free energies cannot easily be obtained from simulations, free energy perturbation methods allow the calculation of their relative values. Often the energy difference must be calculated over a number of intermediate states using a coupling parameter λ to define a mixed potential function. This has the advantage of allowing the simulation of non-physical transformations and has opened up the opportunities to obtain calculated values of many different free energy related properties.

5 Modelling biomolecules

5.1 Introduction

As computer power has increased steadily over recent years so too has the size of the systems which the computational chemist has the potential to investigate. This has opened up many different applications in the field of biological molecules: proteins, DNA, lipids and carbohydrates. The importance of structure for the understanding of biological systems is exemplified by Watson and Crick's discovery of the structure of DNA (Fig. 5.1). Immediately it could be suggested how DNA might replicate – perhaps one of the most significant scientific discoveries of the century – and in turn this spawned the new discipline of molecular biology. The first protein crystal structures, while almost baffling in their complexity, would provide new insights into the structural basis of their activity; so too, the first enzyme structure (lysozyme) allowed the molecular basis of enzymic catalysis to be viewed. Although technology has made experimental structure determination less difficult than in the past, the process of carrying out an x-ray diffraction study on a protein is still measured in years rather than months. Therefore, it is here that computational methods have a part to play by learning from the precedents of existing biological structures and to utilize this information to make valid predictions of unknown structures.

In this chapter we will concentrate on methods for modelling protein structures as much more work has been done in this area. Most of the interesting studies on DNA, carbohydrates and lipids have relied heavily on molecular dynamics methods. The interested reader is recommended to follow this up using the Further Reading list.

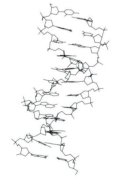

Fig. 5.1 The B-DNA double helix.

5.2 Protein structure prediction

Before continuing, a brief introduction to protein structure is required. Proteins are polymeric macromolecules synthesized from the twenty naturally occurring amino acids which are coded for by DNA. Although other amino acids sometimes appear in proteins, these have arisen from modification after the protein has been synthesized. Fortunately it is possible to create broad groupings for the amino acids according to the properties of their sidechains: hydrophobic, polar but uncharged, polar acidic, polar basic, and finally, glycine, which has no sidechain. The particular sequence of amino acids in a given protein is its primary structure. Note, however, that structure in this case implies nothing about conformation. The next level of protein structure is secondary structure, as defined by the local conformation of the protein backbone. Secondary structure can be further divided into regular secondary structure – such as helices, β-sheets and turns – and irregular structure: loops containing no discernible underlying order. The packing together of these secondary structure units into compact globular domains defines the tertiary fold, while multiple domain proteins require the packing of individual domains to form the quaternary structure. It is important to note that the latter two levels of structure often utilise interactions between amino acid residues which are distant in the sequence.

The recent, vast increase in both protein sequence and three-dimensional structural data has made it increasingly clear that there may only be a relatively small number (1000–2000) of structures formed by proteins. This should reduce the scale of the modelling problem. There are two main strands to this argument. The first comes from the observation that a certain proportion of newly sequenced genes are related to sequences which were already known in the protein sequence database. This relationship is more strictly defined as homology: that is to say, there is evidence to suggest that the proteins have evolved from an ancestral common gene. Coupled with the sequence data is the observation that approximately one third of newly determined three-dimensional structures have a similar fold to previously known structures. Assuming that neither the sequence nor structure databases are biased (not strictly true) it is possible to arrive at the predicted number of ~1200 different protein folds.

The second approach is genetic. Some proteins, which are considered as modern on the evolutionary timescale, are constructed from individual folding units, or modules, which are replicated in a diverse range of sequences; sometimes as multiple copies. These proteins provide some evidence for a process of splicing together gene fragments (exons) to produce new proteins: exon shuffling. That proteins might only fold in a limited number of ways is seductive to the nascent protein modeller.

The final important idea regarding the relationship between sequences and structure is that all the necessary information for the definition of a protein's three-dimensional structure is encoded in its sequence. Crucial to the development of this idea were the seminal experiments of Anfinsen who showed that the protein ribonuclease could regain its native structure when it was removed from unfolding conditions and placed in a solvent which promotes folding. This sets out quite clearly one of the most fundamental unanswered questions in biology: if the three-dimensional structure of a protein is determined entirely by its sequence, what are the rules which map one onto the other?

Protein folding

Conceptually, the most obvious approach to the protein structure problem would appear to be the simulation of the actual process of protein folding. Not only would this have the potential to predict structures from sequences, but by monitoring the development of order throughout the simulation the underlying driving forces, and the structures of intermediates along the folding pathway, could be elucidated. The timescale for the folding event is of the order of milliseconds, therefore it falls outside the capabilities of conventional molecular dynamics simulations using potential functions of

A detailed discussion of techniques used in protein folding simulations is given in Skolnick, J. and Kosinski, A. (1989). *Ann. Rev. Phys. Chem.*, **40**, 207–235.

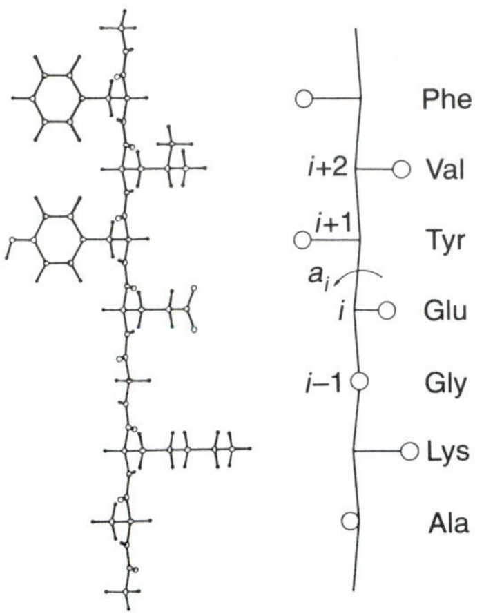

the type discussed in Chapter 3. Therefore, less detailed models must be investigated.

Simplicity can be engineered by two approaches: the representation of the amino acids, and the degrees of freedom allowed to the folding chain. Like many approaches in the protein modelling field the reduced representation of the amino acids can be calibrated using the Brookhaven database of known structures. At the core of a typical protein the principal ordering force is that of close-packing of residue sidechains; at this point we will assume that the protein chain has undergone collapse from its denatured state to maximize burial of hydrophobic residues. Therefore, packing considerations can be reduced to one of sidechain size: the residue can be considered as a volume unit. By examining the occurrence of a given amino acid in known structures one can obtain average values for this volume so the protein is now a flexible backbone with sidechains approximated by spheres of appropriate size. Further simplification is possible by considering the backbone as bonds between C_α (see Fig. 5.2) rather than as two rotatable bonds (ϕ, ψ) and the fixed peptide link (ω). As volume is the only criterion in this representation a potential function of the Lennard-Jones type would be suitable. Models of this type can then be used with molecular dynamics and Monte Carlo methods.

Fig. 5.2 All atom and simplified representation of a peptide. Rotation about torsion a_i links the four residues: $i - 1$, i, $i + 1$ and $i + 2$.

The second approach commonly used is to model the structure as a lattice. The individual amino acids in the chain occupy points on this lattice and change position by movements from one grid point to another. While studies have been carried out with both diamond and cubic lattices, the latter has proved to be the more successful in that it allows a better representation of the secondary structure. Again, a reduced representation of the individual residues is necessary; note that the backbone torsion angles have very little meaning since they do not drive the changes in conformation. The accuracy of this type of model is determined both by the resolution of the grid and the type of potential function used to calculate the interaction energies. Some of the most successful studies have been carried out on a face-centred cubic lattice with point separations of 1Å. If the protein is mapped onto this lattice by placing each C_α atom on a lattice site and considering the inter-C_α vectors to be (2,1,0) the overall topology of the protein is not unduly distorted. But with more crude lattices the structure loses this accuracy of representation and becomes more like a model of a folding chain with some protein-like properties. If an additional atom centre is defined for each residue to represent the sidechain, the lattice model can be modified to include rotational flexibility about the C_α–C_α bonds. A lattice type model is very suited to Monte Carlo simulations. Starting from random chain conformations on the lattice the protein can be folded by moves which change the position of a residue from one lattice point to another; if the sidechain is also represented, random torsional moves about the C_α–C_α bonds will also be required to reorientate its position. Then the energy of the new conformation can be evaluated and the move accepted or rejected using the criteria discussed in Chapter 4. A commonly used method to evaluate the interaction energies is based on the frequency of particular inter-residue contacts in known structures. Clearly this should favour the formation of a compact hydrophobic core but will be less well defined for residues found on the surface which have few contacts with other groups. Also, since residues on the surface of a protein can be present more for purposes of recognition than for structure determination, their influence is ambiguous.

It is still not clear whether protein folding simulations provide useful information about the nature of the folding process or if they can be utilized to predict the three-dimensional structure of a given sequence. Results obtained experimentally on protein folding are also still open to interpretation. Each of the approximations in the method gives rise to problems: the lattice model is at too low resolution for actual structure prediction while the simplified potential functions used rely heavily on known structures opening up the possibility of inbuilt bias. This is not to say that they will have no use in the future once more is known about the rules behind the protein structure code.

Secondary structure prediction

Prediction of the overall fold of a protein requires the consideration of not just local interactions between residues sequentially close on the chain but also non-local interactions through space. In theory then it should be less difficult to predict secondary structure of proteins than to predict the tertiary fold. Typically, secondary structure can be thought to fall into three or four groups: helices, β-strands, and loops. But the latter may sometimes be treated as turns, which are considered to be regular structure. Returning to Anfinsen's experiments on refolding; a naive interpretation of folding would be that the information to define regular secondary structure resides in the local sequence and that the fold, or tertiary structure, arises from the docking together of different folding elements in such a way that they form a hydrophobic core and place hydrophilic residues on the surface. There are obvious thermodynamic arguments against this hypothesis but if we modify it slightly we reach the conclusion that certain residues will bias the formation of a particular class of secondary structure. It was on this basis that the early methods of secondary structure were derived.

In the 1970s only a very few (< 40) protein crystal structures had been solved but it was already clear that certain residues were associated with different elements of secondary structure e.g., glycines are very common in turn regions, and branched sidechains are disfavoured in helices. The first commonly used method was that of Chou and Fasman which was a literal interpretation of these observations. From an analysis of the known structures they produced propensity tables which showed the probability of a given residue being in a particular secondary structure type. By examining the appearance of strings of residues with a particular preference it became possible to predict the occurrence of secondary structure elements, however, some features of the method require more subjective decisions and this makes it difficult to implement computationally. An extension of the statistical method, by Garnier, Osguthorpe and Robson (GOR), considered relationships between residues in the different classes of secondary structure. Interactions with up to eight residues on either side of a given position are summed to give the probability of finding a given residue type in a particular

See, for example, Fasman, G. (ed.) (1989). *Prediction of protein structure and principles of protein conformation*. Plenum, New York.

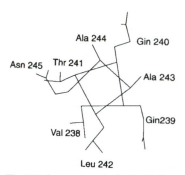

Ala 244 Gln 240

Asn 245 Thr 241

Ala 243

Gln239

Val 238

Leu 242

Fig. 5.3 Arrangement of sidechains in an α helix. The helix is viewed along its helical axis from the C-terminal end.

Sequence alignment methods are extensively reviewed in Barton, G. J. (1990). *Methods in Enzymol.*, **183**, 403–428.

conformation. If we consider a helix (Fig. 5.3) then it is obvious that the sidechain has a greater interaction with those residues three and four positions distant than with those immediately beside it. Thus there would seem to be some justification for including the preferences of other residues. Although more automatic in its use, the GOR method still produced predictions of about 60% accuracy, not much better than Chou-Fasman. So, while statistical considerations of residue preferences can give an indication of some of the secondary structure they also highlight the problem that tertiary interactions can influence structure formation to quite an extent.

Before going on to discuss more recent methods of structure prediction we must digress and outline methods to obtain optimal alignments between protein sequences. This is required for nearly all of the methods still to be discussed.

Sequence alignment

When a new gene product has been sequenced, one of the first things to be done is to check it against the database of known sequences to see if there is any similarity. In this way information can be obtained regarding the classification of the protein and possibly an indication of the structure, or at least important residues. A vast number of sequence alignment tools have been developed to carry out efficiently this kind of screening but we shall consider only the principles behind the general method of dynamic programming.

Obviously sequence alignments can be carried out 'by hand' and it may sometimes be necessary to make manual changes to the automatic alignment, but for mass screening the automatic method is advisable. If we take two sequences, A and B, and assume there is some similarity between them, the problem becomes one of obtaining the best match between the identical residues while pairing non-identical residues in such a way that some common property is conserved as far as is possible. The starting point in all dynamic programming is to construct a two-dimensional matrix where the elements represent the similarity between any residue in sequence A with each of the residues in sequence B. Figure 5.4a shows the simplest case where a value of one is assigned if the residues are identical and zero if not. Next we must find the highest scoring path through the comparison matrix to obtain the optimum alignment. This is known as generating the maximum match matrix. Now consider what a matrix element (M_{ij}) in the maximum match matrix represents: it is the highest scoring path which can be obtained starting at that point. Clearly, since we are restricted to retaining the sequence order, the matches which follow positions i and j can arise only from higher values of the subscripts. In practice we start from the bottom right hand corner and systematically go through the comparison matrix converting it into the maximum match matrix by finding the highest value in the row and column i + 1, j + 1 of the maximum match matrix and adding this to the value in the comparison matrix. The partially completed matrix in Fig. 5.4b shows which values are considered at a given point. Having completed the assembly of the maximum match matrix one starts at the top left hand corner and finds the optimum alignment by choosing the highest value in each row and column pair such that i and j are both higher than those of the previous pair (Fig. 5.4c). For the two similar sequences in the example the highest value commonly fell on the diagonal, but occasionally a letter is skipped, leaving it unpaired. Hence, gaps can be created in the alignment. In evolutionary terms this is justifiable as additional bases can be inserted into DNA as two sequences diverge from a common point. Therefore, it is not necessary to obtain an exact match. However, the creation of gaps must be controlled, otherwise, if dissimilar stretches of residues crop up in matched sequences the algorithm will react by increasing the length of the gaps. To overcome this problem gap-penalties are introduced. These are subtracted from the maximum match score to disfavour moving off the diagonal when different residue types are matched.

The second issue in sequence alignment is the values used in the comparison matrix. In the example, we used the identity matrix. Many different relationships have been constructed to group residues by similarity or by their acceptability as mutations in homologous proteins. Take, for example, the hydrophobic residue leucine; if this were matched against isoleucine or valine the property of the sidechain at that site in a protein would be conserved therefore it would not be an unwise alignment. Thus, some account must be taken of conservative substitutions which would be counted as zero by the identity matrix.

Having screened a database using pairwise alignments, as described above, and obtained a family of sequences which show evidence of being related, the next task is to try and optimally align all the sequences. We will see later on that this provides much more information than a single alignment. Clearly the dynamic programming method could be extended into many dimensions for simultaneous alignment but this would soon run into problems with memory requirements. An alternative approach is first to generate pairwise alignments between all the members of the family. The most similar pair is then taken as the core of the alignment to which successive sequences are matched, with average values being placed in the comparison matrix where there is not identity across the growing set of alignments. This method can be run iteratively to find a self-consistent solution. The main consequence of a multiple alignment is that positions in the sequence which show absolute conservation, or where only conservative changes are made, can be spotted.

Comparison of known, related three-dimensional structures have shown that the rate of successful mutation is much greater on the surface of a protein than it is in the protein's core. Thus by monitoring the degree of conservation across a family of proteins we can start to make predictions about tertiary structure without prior knowledge of the three-dimensional structure.

Methods utilizing this approach have recently been used to great effect for secondary structure prediction prior to the publication of the experimental results. The methods can be summarised as follows:

1. Collect an appropriate set of sequences and obtain an optimal multiple sequence alignment.

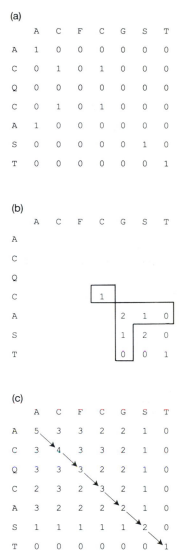

Fig. 5.4 The main stages in sequence alignment. (a) Construction of the comparison matrix. (b) Conversion of the comparison matrix into the maximum match matrix. The value in the highlighted position is added to the highest value in the boxed region. (c) The completed maximum match matrix. The final alignment is read from the top left hand corner; in this case it is along the diagonal.

2. Distinguish between buried and surface regions using residue variability and patterns of hydrophobic residues.

3. Look for key residues, such as glycine and proline, which tend to break regular secondary structure regions; the sequences are now divided up into likely regular and irregular secondary structure regions.

4. Assign secondary structure to the regular regions. The Benner method uses the repeat of property patterns to make this assignment e.g., a β-strand has alternate residue sidechains above and below the plane of the β-sheet; a surface β-strand would therefore have alternating hydrophobic and hydrophilic sidechains. Similar arguments can be made for helices.

5. Attempt to construct a tertiary structure template using known biochemical information such as the active site residues, epitope mapping, and site-directed mutagenesis data.

See Benner, S. A. (1992). *Curr. Opin. Struct. Biol.*, **2**, 402–412.

This last stage is the most difficult to achieve, but even its exclusion leaves a potentially more powerful method than those discussed previously. Note, however, that the structure produced is not that of an individual sequence but a consensus structure which provides a representative template for the complete family of proteins.

The inverse folding problem

At the start of the chapter it was implied that, although the three-dimensional structures of a great proportion of sequences are not known, in many cases the sequence might be expected to assume a fold similar to that of a known structure. From sequence alignment one can pick up closely related sequences, with maybe > 40–50% sequence identity. But how do we find more divergent relatives? This can be rephrased such that if we have a known protein fold, how do we find out which other sequences are consistent with this structural pattern? Thus described we have the 'inverse folding problem'.

Substitution tables: Overington, J., Johnson, M. S., Sali, A., and Blundell, T. L. (1990). *Proc. Roy. Soc., Lond. B*, **241**, 132–145. 3D–1D Profiles: Eisenberg, D., Bowie, J. U., Luthy, R., and Choe, C. (1992). *Faraday Discussions*, **93**, 25–34.

The starting point for this type of study is an examination of the database to extract data which groups residues as similar according to their properties. If one starts with families of homologous proteins whose structures are known, structure alignment can be used to superimpose the folded chains giving equivalent positions in the sequence. These equivalencies might not be accurately determined by sequence alignment due to the nature of the method, but when using three-dimensional structures they are unambiguous. Comparisons between the residues found at a particular position across all the equivalent sites in the family give an indication of which substitutions are acceptable and which would be disfavoured. To relate this information more closely to structure it is necessary to create a set of property classes utilizing information about whether residues are buried or on the surface; secondary structure classes; positive ϕ angles in the backbone (indicative of turn regions), and hydrogen bonding properties. Classifying all of the substitution data as above produces tables which are essentially probabilities for replacing one residue by another at a site in a protein with given properties. By bringing all these data together into a master table one can use the method for conventional sequence alignment whereby one structure is known, with the properties at all the positions calculated, and the second structure is matched using the substitution tables to complete the comparison matrix. Thus the alignment shows the probability of the unknown structure adopting the fold of the known structure; an alignment with high scoring matches and the minimum of gaps would appear to suggest similarity, but the converse situation would imply that the two were not related. Another interesting property of this method is that apparently anomalous results can point to interesting features such as buried polar residues (usually disfavoured) which could be involved in, say, the active site of an enzyme.

The second method, known as 3D–1D Profiles, uses a far smaller number of structural classes: eighteen. These are derived from three levels of sidechain burial; three classes of secondary structure, and two classes defining the amount of buried surface in contact with polar residues. Thus a starting protein structure is converted from a sequence of amino acids to a sequence of residue environments. Propensities for a given residue existing in each environment are calculated from an analysis of the composition of known protein structures. These propensities can now be used in sequence alignment methods to determine whether sequences of unknown structure would be consistent with a particular fold. The advantage of 3D–1D Profiles and the previously discussed substitution tables is that once the propensity values are known the method can be used for screening sequence databases to find likely candidates to assume a particular three-dimensional structure. In both cases the secondary structure of the starting structure is known so high gap-penalties can be used to disfavour the breaking of these regions in the alignment. Also, one is no longer restricted to arguments based on homology since one is interested only in consistency of sequence with fold, not that the two sequences are necessarily descended from a common ancestor. This opens up the possibility of finding homologous matches with low similarity scores and even structures which have arisen from convergent evolution but which may have no sequence similarity whatsoever. Obviously the propensities are only as good as their statistical significance but as more structures become available it should be possible to recalculate these values to reflect the extended database.

One interesting extension of the 3D–1D Profile method is its use to check experimental and model structures. If one takes a set of protein coordinates and calculates the environment at each residue site, the score for the native sequence in that environment can be plotted against residue number. Significant dips in this profile can point out residues in unsuitable environments and in some cases this has been shown to result from mistraced chains in the crystal structure analysis. Figure 5.5 shows the profiles for myoglobin and a myoglobin structure containing a deliberately misfolded region. Clearly this is a potentially very powerful tool for the verification of model built structures.

The final method to be discussed in this section approaches the problem from a different angle in that instead of posing the question 'which other sequences are likely to fold in the same way as protein X?', it asks 'which three-dimensional fold is the most likely to be assumed by sequence Y?' It is also known as 'threading'. The authors have created a database of unique

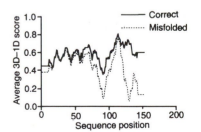

Fig. 5.5 Comparison of the 3D–1D profiles for the correct structure of myoglobin with a deliberately misfolded structure. The incorrectly folded regions are obvious by their lower scores.

protein folds against which sequences can be evaluated. The sequences are matched using a dynamic programming method which places energy values in the comparison matrix. The energy values are calculated using a potential of mean force which is, in essence, a distance-dependent probability of finding a given pair of residues. An additional solvation term has been added for surface residues. The advantage of the dynamic programming method is that it allows the systematic investigation of the effect of placing gaps in the sequence by way of a penalty term in the energy function. Thus the comparison matrix can be converted to the maximum match matrix giving the energetically most favourable match of the sequence to the structure. And by carrying out the process across the full range of known folds the most acceptable structure can be found. Again, this method can pick out non-homologous structures, or domains which might be similar in fold. Also, subject to the limitations of the potential function, it has a greater chance of picking up more marked changes in sequence, such as those resulting from switches in polarity, since the sequence of the three-dimensional fold plays no part in the threading procedure. This is not the case for the other two methods.

Therefore, we are now in a position to generate good models for sequences which can show evidence of assuming a fold close to those which are known and can pick out important residues for defining membership of a particular family. In addition, we can identify distantly related or convergently evolved structures. The next step is to use this information to generate model structures.

5.3 Modelling by homology

The discussion in the preceding sections has shown how sequence alignment and other methods can identify structures which are potentially similar. This can now be exploited to produce three-dimensional structures of proteins prior to their experimental elucidation. However, one should always be cautious that the relationship between the unknown structure and the sequences of the known structures is significant. At the outset it should be made clear that there may be errors introduced into the model structure which arise from the approximations of the method. This is often used to criticize the utility of the method, but as structure-based methods of drug design become a more prominent tool in the pharmaceutical industry, and the desire to provide a molecular basis of biochemical processes becomes more pressing, it is better to have a carefully built model structure than no structure at all.

First, some terms will be defined. The generic name for the method is homology modelling, where homology is strictly defined as meaning descended from a common ancestor; there are no degrees of homology, as is sometimes quoted in the literature: a protein either is or is not homologous to another protein. At the time that the method was being developed only structures which had a high level of sequence similarity could be easily identified as candidates for the method so it was limited to homologous structures. Now, as methods such as 3D–1D Profiles and threading allow the

For further details see Blundell, T. L., Carney, D., Gardner, S. Hayes, F., Howlin, B., Hubbard, T. *et al.* (1988). *Eur. J. Biochem.*, **172**, 513–520.

identification of possibly convergently evolved structures, the strict ancestral link has been broken; we are no longer doing homology modelling, but the name has stuck. Other important terms are Structurally Conserved Regions (SCRs) and Variable Regions (VRs). The former are those parts of the known protein or proteins which show conservation between one structure and another; typically these will be the basis of the core or framework of the protein and will include most of the regular secondary structure regions. The variable regions are stretches of the protein where there is more sequence variability and, often, conformational differences. Fortunately, these regions of the protein will be found on the surface of the protein and the gaps indicate either a reduction or extension of turns or loops. Figure 5.6 shows a flow diagram outlining the different steps in the modelling procedure. These will now be considered in turn.

The alignment

It is usual for a newly sequenced protein to be screened against the database of known sequences to determine whether it shows any similarity to a particular class or family. If one restricts this database to those sequences whose three-dimensional structure is known, a successful alignment will produce one or more suitable candidates from which to build a framework. Alternatively, inverse structure prediction methods will give an alignment of sequence to structure without the requirement of sequence similarity. Studies have shown this stage to be the most important of the whole modelling process so care must be taken to check that the alignment has done nothing to break the integrity of the starting structure i.e., introducing gaps in regular secondary structure; burial of charged residues etc. A sample fragment of an alignment is shown in Fig. 5.7.

If a family of structures has been selected as potential templates it is necessary to carry out a multiple sequence alignment to obtain the maximum amount of information. This will give a clearer indication of the more conserved and variable regions in a typical member of the family and also highlights residues which are invariant across the complete set of matches. These should include active site residues, disulphide links if present, and conserved co-factor binding sites. Conserved residues are also important anchor points for the SCRs.

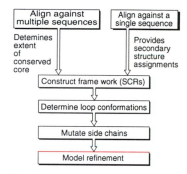

Fig. 5.6 Flow diagram of typical steps in homology modelling.

Construction of the framework

When only one structure is being used as the template for modelling, this stage in the process reduces to a manual inspection of the structure to determine the limits of the conserved regions. For example, if we require to make an insertion into a loop between two regions of β-strand (Fig. 5.8), it would be sensible to replace the whole loop by a longer one rather than attempt to incorporate additional residues into the existing one. In this case the SCRs are the two β-strands. Clearly, when using only one structure, each insertion or deletion must be viewed on its own and there is scope for subjective judgement. If multiple structures are being used this is not the case.

Sibanda, B. L., Blundell, T. L., and Thornton, J. M. (1989). *J. Mol. Biol.*, **206**, 759–777.

```
TP: G I Q V R S G Q D N I - - - N V V
C2: G - - - - N D H S L W R V N V G
FB: D D - - - - K E H S I - K V S V G
```

Fig. 5.7 Fragment of a multiple sequence alignment using the serine proteases trypsin (TP), C2 and factor B (FB). Gaps in the alignment are indicated by dashes.

From the multiple sequence alignment we hope to have obtained a set of key invariant residues. These can now be used to superimpose the structures of the templates in such a way that there is optimum matching of regular secondary structure. The SCR is then defined by starting from the invariant residue and moving along the aligned structures, position by position, on either side, calculating the root mean-squared deviation of the residue positions across the multiple structures. If the structures are close, the SCR is extended by one residue in that direction. This procedure is continued until the chains start to diverge i.e., the variable regions are reached. In this way a set of SCRs can be constructed to represent the framework of the protein. One can either take the average coordinates across the family of structures or choose the structure with the highest degree of sequence similarity for the model SCR. Figure 5.9 shows the SCRs from a set of aligned structures.

Selecting variable regions

If the preceding steps have been carried out carefully the remaining regions to be modelled, the variable regions, should be loops or turns on the protein's surface. A number of different approaches can be taken here and it is suggested that a combination of these methods, rather than dogmatic adherence to one, be used. Usually, modelling the variable regions will require some kind of insertions or deletions in the chain but this must be done in such a way as to minimize distortion of the existing framework.

The first case to be treated is turns linking antiparallel stretches of secondary structure, most commonly β-strands. Usually the minimum number of residues are involved and the conformational restrictions are quite considerable. To make a change to this type of structure it is advised that the tables of different β-turn geometries compiled by Janet Thornton and co-workers be consulted. These have been derived from an analysis of all such turns in the protein database and point out conformational preferences for a given length of chain. Also, their classification incorporates information about the hydrogen bonding pattern on either side of the turn so appropriate structures can be picked which do not change the register of the chain following the loop.

In many cases the loop structure to be modelled is less strictly defined than above. These can be approached in two ways: database searching or conformational searching. The first of these methods uses the protein database as a source of 'spare parts' for protein modelling. If we require to make an insertion into the loop shown in Fig. 5.8 our starting information is the length of the loop, its sequence, and the geometry of the framework regions on either side. We can then search through the database looking for a chain of the required length but subject to the constraint that the neighbouring framework must be close to that of our template structure. Most commonly, the matrix of inter-C_α distances is used to compare the framework geometries but additional information such as the backbone dihedral angles at the junction points may also be usefully compared; if these differ markedly to the original structure, the direction that the new chain takes would be quite different. A second indicator could be the mass-centre of the loop. By using these criteria we are minimizing the changes made in the modelling procedure.

The second method is conformational searching. As discussed in Chapter 3, this method, while being rigorous, suffers from the sheer number of conformations to be generated as the size of the loop increases. Unless one has access to hugely parallel computers and carefully constructed, efficient algorithms this method cannot realistically be useful for generating a loop of more than seven residues. Since amino acids can only exist in certain broadly defined conformations the search procedure can be reduced to combinations of these but computer power can still be the limiting factor. Once generated, a set of suitable conformations can be assessed by the same criteria as those located in the database search.

Finally, it is worth mentioning immunoglobulin loops. These are of particular importance since they provide immunoglobulins with their diversity of recognition powers. However, from an analysis of the known three-dimensional structures it was shown that five of the six loops existed in only a small number of conformations, since called the canonical structures. More recent experimental evidence has validated this concept with almost all of the new structures conforming to these rules. One non-canonical loop structure has been found but given the size of the original sample of structures it is surprising that so few exceptions have turned up. Studies have pointed out which residues in a particular loop in an antibody are responsible for determining the conformation, so now most of the loops in an antibody can be modelled using the most suitable canonical structure. The final loop, for which no consensus has been found, must be modelled using other searching methods.

Sidechain placement

Studies on the conformations of sidechains in homologous proteins have shown conservation as far as is possible. Therefore, the first group of residues, those which are identical, is easily treated: no change is made. If a residue substitution is being made then, as far as possible, any common sidechain dihedral angles between the old and new sidechains should be conserved. Again, there is experimental evidence that this should be the case. If one considers the protein as a whole, then the sidechains which are constrained more are those on the inside. Packing considerations will play as much of a role as the conformational energies of the individual residues. Therefore, if a mutation is to be acceptable it must make the minimum possible perturbation to the packing arrangement. Thus the conformation must be conserved.

If the new sidechain is very much bigger one must either look to the database for precedents or search for the most favourable conformation. One interesting approach is to devise rules for the likely conformation on substituting one residue for any other. If this is further refined by including secondary structure classes, a set of 1200 substitution rules is obtained. This method can be calibrated using families of homologous proteins.

Conformational search procedures can be used if the number of rotamers is limited only to those likely to exist i.e., the minima of the torsional

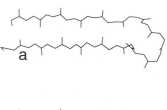
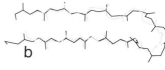
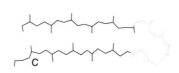

Fig. 5.8 Grafting a new turn structure between two β-strands.

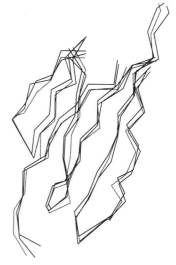

Fig. 5.9 Superposition of three immunoglobulin structures showing the conserved β-sheet framework (SCRs).

potential, and sampling can be restricted to, say, dihedral angles beyond those found in the original residue. In the case where a very divergent structure is being built, or a convergently evolved structure, the number of residue substitutions may be great and retention of original conformation might introduce unwanted strain. An interesting method which has proved very successful is to apply simulated annealing to the sidechain dihedrals within the framework of a constrained backbone. This can produce an optimum packing arrangement of the sidechains in close agreement with experiment.

Fortunately, many of the methods described for homology modelling are available in software packages, but one should be cautious when applying a complete procedure blindly. Structures should be checked closely to avoid errors.

Refinement

The structure obtained from the above modelling steps is likely to be in a very strained state. There will probably be many less than optimal non-bonded contacts, together with deviations in bond lengths and angles at the joins between SCRs and variable regions. Therefore, the structure must be refined in some way. At this stage it is important that the changes in the structure are annealed with as little modification to the starting framework as possible. To do this we must use a gradual approach and try to, first, refine poorly defined regions with respect to a constrained framework, then release the constraints and allow further relaxation of the structure. It is important that the final refinement is done using as few constraints as possible to minimize any bias in the procedure.

Refinement can be carried out using just energy minimization but, if packing of the protein's core is to be achieved, it will probably be necessary to use molecular dynamics to avoid the structure being trapped in a local potential well.

The degree of correctness can be assessed in different ways. First, one should check that the structure fulfils the criteria normally expected of an experimental structure: it should show good stereochemistry and not have residues in the disallowed regions of the Ramachandran map (Fig. 5.10). Second, any residues in unusual environments should be investigated. These will sometimes be apparent by eye but the 3D–1D Profile method of Eisenberg should give a more systematic picture. Third, check that features such as the solvent accessible surface area are typical for a protein of the length of your sequence. This is a much more sensitive property to the correctness of folding than the molecular mechanics energy. Finally, use any available experimental data to check that these observations are compatible with the model. Only at that stage can one be sure that the model is, in essence, correct.

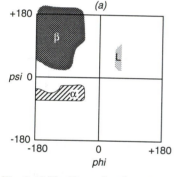

Fig. 5.10 The Ramachandran map showing the allowed conformational regions for amino acids. The highlighted region indicate α-helical (α), β-strand (β) and left handed helical (L) conformations.

5.4 Modelling enzyme reactions

Although much of the modelling work on biological molecules has focused on structural aspects there is now increasing interest in methods which can successfully simulate enzyme reactions. The kind of information which is produced experimentally – binding constants, rates of reaction – provide little feeling for the molecular basis of catalysis and cannot prove that a particular mechanism is followed. This would appear to be an ideal place for computational methods to fill in the gaps left by the experimentalists and at the same time provide new insights into the way that enzymes work. One of the most important aspects, yet most difficult to quantify, is the nature of the transition state for a given enzymic reaction. It is now generally accepted that the driving force of enzymic reactions results from the stabilization of the transition state compared to what would be expected in solution. The elusive nature of the transition state makes it difficult to describe structurally with the result that vague hypotheses based upon organic chemical intuition, rather than hard structural explanations, exist for most reactions. Without this proper understanding of the nature of a particular transition state we still lack detailed information about why an enzyme works and cannot suggest how its efficiency could be improved.

A number of methods have been used to probe enzymic reaction mechanisms, and this is still a field under of development. Here we shall discuss some of the approximations which have been made to overcome the problems associated with systems of this size.

A general review is given in Mulholland, A. J., Grant, G. H., and Richards, W. G. (1993). *Protein Engineering*, **6**, 133–147.

Small model systems

In an enzymic reaction bonds are made and broken, therefore explicit account must be taken of the electrons in the system. For this reason molecular mechanics methods are of little use. A common approach has been to model the enzyme as only those residues actually participating in the reaction. A further approximation is to treat amino acid residues as just their functional groups: the carboxylate of an aspartate or glutamate sidechain is modelled as formate; histidine by imidazole, and a peptide linkage by formamide (Fig. 5.11). In this greatly reduced form the reacting system can be investigated using *ab initio* or semi-empirical molecular orbital methods (see Chapter 2). Thus the system is treated essentially *in vacuo* without reference to its surroundings. In a typical case the model system will have the reacting groups arranged as they are found in the enzyme's structure. Since the 'bare' reacting groups feel no stabilization from the surroundings, as would be the case in the real system, it could be necessary to restrain their movement. Often the concentration of polar groupings will result in very high attractive or repulsive forces. The system can then either be driven along a reaction coordinate by changing the geometry of the model, providing an indication of the energetic pathway for the reaction, or transition state locating algorithms can be used to find the structure of the transient species and calculate their energies. If one is interested in the specific effect of surrounding residues, the appropriate functionality can be added but again its effect may be greatly exaggerated.

In addition to the acknowledged shortcomings related to the size of system, model calculations using *ab initio* wavefunctions will suffer from all of the usual problems associated with basis set size. More recently a number of successful applications of semi-empirical methods have highlighted this

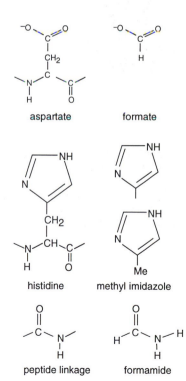

Fig. 5.11 Mimicking amino acid functionality for use in small model systems.

option as having considerable potential; also the size of the system can be greatly increased.

One of the simplest ways to include the effect of the environment into a model enzyme calculation is to surround the explicitly quantum mechanical segment by charges on the atom centres of the remaining atoms. This has the effect of modelling the influence of the electrostatic potential of the enzyme. Many electrostatic effects result not from explicit atom-atom contacts but from the focusing of long range charge distributions. Consequently, their inclusion into the calculation is an important step towards more realistic simulations. The increase in computer time is not that great since the charged atoms are considered as nuclei and are only included in the one-electron integrals, not the more costly two-electron integral evaluations. Unfortunately this is only a partial solution in that it accounts for the electrostatic effect of the surroundings on the reacting atom but not for polarization effects in the other direction. Thus the charge field is a rather static representation. Also no account is taken of van der Waals interactions which can play a large part in favourable binding energies.

Attempts have been made to include reaction field effects (see Chapter 4) but only homogeneous fields can be modelled easily. To model an inhomogeneous reaction field, as would be found in a protein, is still not straightforward, and this still leaves the question of the influence of ions and the solvent. Similarly to the charge field method, van der Waals interactions are neglected.

Combined molecular mechanics/quantum mechanics methods

The methods which would appear to hold the greatest potential are those which treat the central reacting system by explicit quantum mechanical methods while allowing the remainder of the system to be represented by a classical model. The energy of the system therefore becomes

$$E_{\text{tot}} = E_{\text{QM}} + E_{\text{MM}} + E_{\text{QM/MM}} \tag{5.1}$$

where the first two terms E_{QM} and E_{MM} are reasonably straightforward to calculate. For this type of calculation the reacting system is represented by a similar model system to those discussed in the previous section. The linking bonds between the QM and MM are replaced by hydrogens pointing along the direction of the bonds. The non-bonded interactions between atoms in the molecular mechanics segment of the system are calculated as usual. The last term in Eqn. 5.1 represents the interactions between the QM and MM atoms: van der Waals interactions between QM and MM atoms are calculated using the appropriate potential for the force field (Lennard-Jones, Buckingham etc.), while electrostatic interactions are incorporated by considering the MM atoms as notional core s orbitals. These can be used to calculate a charged core-core interaction. This latter formulation is necessary due the nature of the MNDO formalism used to treat the QM atoms; if an *ab initio* method had been used the charges could be included as discussed in the previous section.

Clearly this method has some of the disadvantages associated with molecular mechanics calculations in that the basic MM representation cannot account for polarization effects. However, it has been used very successfully to model the reaction carried out by triose phosphate isomerase. On comparing energetic pathways calculated using protonated and neutral histidine residues it was clear that only the latter pathway was consistent with experiment.

The method is discussed fully in Field, M. J., Bash, P. A., and Karplus, M. (1990). *J. Comput. Chem.*, **11**, 700–733.

The final method to be discussed is the Empirical Valence Bond Method (EVB) due to Warshel. This is distinct from the other methods in that it does not use molecular orbital theory. First, one must divide the system into the reacting part and the surroundings. For the reacting section one must derive all the possible valence-bond resonance forms; these will be both covalent and ionic. This has the advantage of representing the reaction by chemically intuitive forms. The energy of the reactive system is then found by using the secular determinant formed using all of the resonance structures. The pure forms are placed on the diagonal and the off-diagonal terms represent mixing terms. For the covalent forms the appropriate energy term is usually an empirical force field function, while the ionic resonance forms include an electrostatic term. Diagonalization of this determinant produces an analytical form representing the system so that the energy can now be calculated for any given geometry. To ensure a close fit to experiment the calculation is then calibrated using experimental data on representative systems in solution. When the reaction is carried out in the enzyme, the active site is considered as being 'solvated' by the surrounding protein and non-bonded interactions replace the solvation energy contribution. This is usually done using more complex representations than in a typical molecular mechanics method so allowing polarization of the surroundings. The EVB method has been used to investigate a number of systems and from the results it is clear that electrostatic effects play an enormous part in enzymic catalysis. Also, from the relative contributions of the different resonance forms one can discern the nature of the transition state complex.

An extensive description of the EVB method is given in Warshel, A. (1991). *Computer modelling of chemical reactions in enzymes and solution*. Wiley, New York.

5.5 Summary

In many circumstances it is now possible to make a reasonable prediction of the structure of a protein from its sequence. Methods which predict secondary structure will typically be around 60% reliable but, together with information from multiple sequence alignments, it is possible to make quite realistic divisions into regular and irregular secondary structure. Multiple sequence data also allows the identification of the most likely mutations at particular positions across a family of homologous proteins. Extension of this information to include different structural classes provides excellent tools to approach the inverse folding problem. The first step is to consider the three-dimensional structure as a sequence of environments, then consistent sequences for those environments can be found.

Once a sequence of unknown structure has been matched with a known structure, homology modelling methods can be used to effect the transformation of one structure into the other. Since insertions and deletions are found in the loops and turns on the surface, the secondary structure of the original protein remains intact. Again, multiple structures will provide a better consensus.

Theoretical investigations of enzyme mechanisms are now also a possibility. These can range from rather limited studies on gas-phase models up to methods incorporating quantum mechanical fragments into the molecular mechanics force field. Such methods are proving to be very useful in understanding the molecular basis of enzyme activity.

6 Ligand design

6.1 Introduction

One area where computational chemistry has proved to be most useful is in the design of molecules as putative therapeutic agents. The structure of the drug, its receptor, and the interactions between them are analysed closely to find the best way to optimize activity or to give an indication of which molecules would be the most interesting to synthesize. Historically this was not always the case and drugs were discovered by the often discussed phenomenon of serendipity, rather than deliberately designed. The major breakthrough in the development of a rational process of drug design came from the Nobel Prize winning work of Sir James Black, who took as his starting point the working hypothesis that the structure of an inhibitor for the transmitter molecules noradrenalin or histamine should be based upon the structure of the natural transmitter itself. By extending the structure in different ways, and by adding appropriate functionality to the ring and sidechain, he was able to use biological activity to map out the active site in terms of acceptable size and polarity. This systematic approach was the norm until the late 1970s.

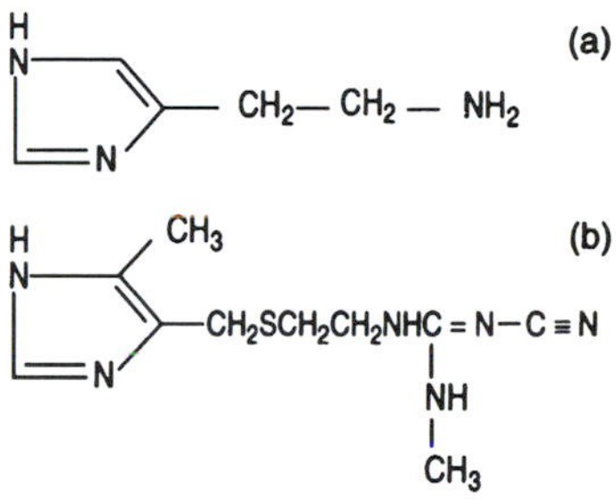

Fig. 6.1 Structures of (a) histamine and (b) cimetidine.

Empirical studies such as the type outlined above were greatly helped by the development of methods for Quantitative Structure Activity Relationships (QSAR). These techniques can be considered as a natural extension of the methods used in organic chemistry for predicting linear free energy relationships, and the Hammett substituent reactivity constants which had been in use for many years. The idea was to take biological activities and relate these to the properties of the molecules. These properties could be related to size (steric), lipophilicity, or electronic features not dissimilar to the Hammett constants. For a group of compounds a regression equation such as Eqn. 6.1 could be fitted to the biological activity, giving an indication of the relative importance of each property to the activity

$$\log(1/C) = k_1 + k_2 \log P + k_3 (\log P)^2 + k_4 \sigma + k_5 E_S \tag{6.1}$$

In an equation of the type shown above the biological activity factor – represented as $\log (1/C)$ – is considered not just from the viewpoint of receptor binding but also in terms of the transport capabilities of the molecule. The latter is characterized by partition coefficient (P) contributions. The σ term is electronic and E_S is a steric term. Thus, while one is attributing activity to the different properties of the molecule there is still some confusion concerning the hydrophobic nature of the molecule as this could contribute both to transport and binding. Therefore, the regression equation could suggest that the drug be optimized in a misleading way such that a poor binding molecule could give reasonable activity because of optimization of transport properties allowing it to be present at the receptor in high enough concentrations to be active. This is clearly the opposite of the desired result but the blame cannot be attributed fully to QSAR analysis: the nature of the pharmacological data often would not allow further discrimination between the different effects. Fortunately, molecular biology has opened up the possibility of using cloned receptors so biological activity can be separated into its two components: transport and receptor affinity. This less ambiguous data could lead to the more fruitful application of QSAR methods in the future.

3D-QSAR

See Cramer, R. D., III, Paterson, D. E., and Bunce, J. D. (1988). *J. Am. Chem. Soc.*, **110**, 5959–5967.

Since the availability of molecular graphics systems has become more widespread, interest has shifted towards methods which are more intimately concerned with the detailed molecular structure of drug molecules; parameters such as partition coefficients are bulk properties. Improvements in computing power have also allowed more thorough conformational analysis on flexible molecules to be carried out. Thus, structures can now be readily predicted and visualized.

The principal problem when considering drug molecules is the conformation actually assumed at the receptor. There is no reason to believe that this will correspond to structures obtained by crystal structure analysis or the solution conformation from NMR. This forces one to think in terms of consensus structural elements across a range of different molecules. Let us first assume that we know how to carry out molecular superpositions successfully; in later sections this will be discussed in more detail. This allows the evaluation of the influence of different structural features on the behaviour of a drug molecule at the binding site: so called 3D-QSAR. First, a grid is placed around the collection of superimposed molecules. For each molecule in the ensemble interaction energies are calculated using a probe placed at each grid point in turn. This interaction energy is composed of a van der Waals term, usually with respect to an sp^3 carbon atom, and an electrostatic term. If these two energies are treated separately the structure-activity relationship can be viewed in terms of the more traditional steric and electrostatic contributions. For simplicity we shall assume a ten point cubic grid. So, for each molecule we have 2×10^3 energy values (steric and electrostatic) and one value for the binding constant. Now, one must assess the importance of each grid point in determining the binding constant; we are attempting to produce a regression equation in terms of the 2000 variables. Obviously this is a vastly underdetermined problem, as in most cases, one will have only tens of compounds against which thousands of coefficients must be fitted. Fortunately modern least-squares methods can handle such cases.

Analysis of such a huge regression equation can best be done graphically: the grid points surrounding the collection of active molecules can be coloured according to the size of the coefficients and very small values can be ignored. In this way clusters of points can be used to identify regions of common steric or electrostatic energy and thus provide an indication of the

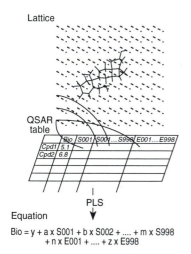

Fig. 6.2 3D-QSAR analysis.

most important regions of the molecules in determining activity. Future molecules may then be designed in such a way that their properties are optimized to these assigned regions of space.

By concentrating on the fields generated by a molecule, rather than the identity of chemical groups, 3D-QSAR can be thought of as seeing the molecule from the viewpoint of the receptor. As is typical in these cases, the parameters used for comparison will always be open to criticism; particularly the charges. Also, the molecular superposition and the size of the grid spacing will be very sensitive parameters. However, in the light of these shortcomings, the extension of QSAR methods beyond the use of substitution constants must be considered as a major advance.

6.2 Inferred receptor sites

Active analogs

Although advances in technology are now allowing the characterization of many protein structures, the problem still remains that many of the targets for drug intervention are membrane-bound receptors about which relatively little is yet known. Medium resolution electron microscopy studies have allowed the gross structural features of these receptors to be visualized, but this is still some way from providing a complete set of coordinates. Therefore, the nature of the receptor site must be inferred from the structures of known binding agents. This has become known as the 'active analog approach'.

Clearly the best starting point when attempting to derive a receptor model from known active agents would be a high affinity molecule with little conformational flexibility. This would serve to define the spatial orientation of important units of the molecule. By important groups we mean functionality which would be likely to make a significant interaction with the receptor on binding i.e., amino groups (but their protonation state must be known); hydrogen-bond donors and acceptors; bulky hydrophobic groups (indicative of a hydrophobic pocket at the binding site), etc. If the set of active molecules contains a rigid structure one immediately has a template onto which the other molecules may be fitted. When all the molecules contain some conformational flexibility there are two options: either one can assume that the lowest energy conformer in each case will be the binding conformation, and then attempt to superimpose these structures, or alternatively, represent each molecule by an ensemble of conformations below a certain energy threshold and choose the structure which gives the best fit across the set. The latter is obviously much more time consuming in that many more flexible superpositions must be considered, but makes fewer assumptions about the influence of the receptor on ligand conformation. If only low energy conformations are used, one runs the risk of being completely wrong, or at least of overlooking possible better matches. If the molecules can be fitted using conformations which are energetically not too disfavoured, it is reasonable to assume that the energy difference can be compensated for through interactions with the receptor.

How best can molecules be fitted? If the group of molecules is not too diverse in structure it may be possible that the appropriate matching units are easily spotted and the superposition carried out manually at the graphics screen. Following this rough matching, paired atoms/groups can be better superimposed using a simple least-squares minimization. If it is not immediately obvious how to match structures i.e., there are considerable differences in the exact functionality, it must be assumed that molecular properties such as the electrostatic potential at the van der Waals surface are similar. The electrostatic potential at a point is simply the electrostatic energy of the interaction between all the charges in the molecule and a hydrogen ion at that position. Thus it is a much better indicator of a molecule's polarity than the actual charges on the atoms. Although it may now be obvious as to which parts of each molecule should be superimposed, there could still be problems associated with the irregular shape of each molecule's surface. For this reason some methods have been developed where the electrostatic potential, or some other characteristic property, is projected onto a regular geometric shape such as an icosahedron or a sphere. Thus the problem of superposition becomes one of matching electrostatic potentials on a few relatively evenly spaced points, giving much greater scope for automating the complete procedure. More complex methods such as Molecular Similarity will be discussed in more detail below.

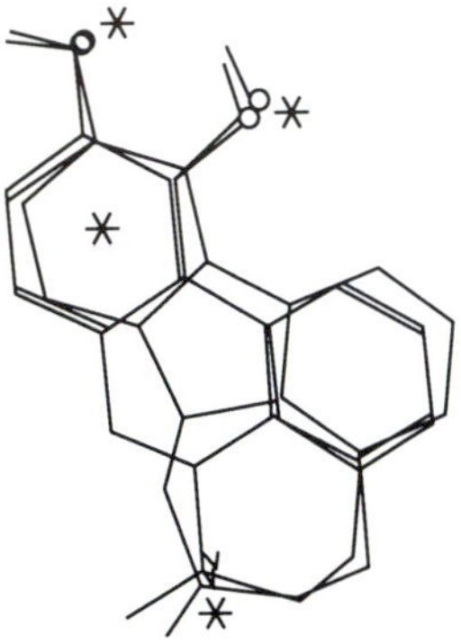

Fig. 6.3 A proposed CNS pharmacophore. The starred groups are likely to be important for binding.

Pharmacophores

Having obtained a set of optimally aligned drug molecules, two different pieces of information can be gleaned: the pharmacophore, and an indication of the size of the receptor binding site. The first of these, the pharmacophore, is defined in each case by extracting common structural elements from the superimposed structures. These will most likely be involved in binding. Each of these sites on the drug molecule is considered to be essential for activity and often may have already been identified as the matching points in the superposition. Figure 6.3 shows typical pharmacophore elements for CNS (central nervous system) active drugs and identifies the optimal distance between the amine group and the centre of the aromatic ring. Also, the angle between the lone pair and the vector normal to the ring-plane can only be within certain limits. If, for example, the drug is directed against a metal-containing active site, then acidic groups capable of binding to the metal should be an important component of the pharmacophore. But each system is different and must be considered separately.

Further discussion of pharmacophores and the active-analog approach can be found in Marshall, G. R. (1987). *Ann. Rev. Pharmacol. Toxicol.*, **27**, 193–213.

Receptor volumes

The second piece of information, the receptor volume, is obtained by application of the active-analog approach. First one requires the superposition of a diverse set of active molecules. If this can be done, and a pharmacophore is apparent, the assumption can be made that each of the drug molecules will bind to the same site on the receptor and so the accessible volume at the binding site must be capable of accommodating the total van der Waals volume enclosing the complete set of molecules. One then does something similar with a set of related molecules which are known to be inactive. Assuming that the same units have been used to superimpose the two sets of molecules it should then be possible to superimpose the two total volumes by matching the pharmacophore elements. Subtraction of one volume from the other will indicate those regions occupied solely by the active molecules. This provides information on the steric constraints which might operate at the active site and so give an indication of how to extend a given molecular structure to optimize its binding. Conversely, volumes occupied solely by the inactive molecules might imply clashes with the receptor and hence modifications in this direction would be disfavoured. Methods such as the active-analog approach are greatly enhanced by the use of computer graphics to visualize the occupied and disallowed volumes and provide an immediate 'assay' for any new structure prior to synthesis.

The most obvious flaw in this method is the assumption that all molecules bind at the same site. Also, one cannot be sure of the reasons for a molecule's inactivity; it may be something other than steric. Finally, a great deal of information is ignored in that one considers only the components of the pharmacophore, and the total volume; very little can be deduced about more subtle interactions between other parts of the molecule and the receptor. However, in spite of these shortcomings, the method has proved to be a very useful starting point for many industrial drug discovery programmes.

6.3 Receptor mapping

The most fortunate circumstances from which to initiate drug design studies are those where a detailed structure of the receptor is known and available. This can be either an experimental structure, from crystallography or NMR, or a modelled structure derived using the methods outlined in the previous chapter. Usually the experimental structures are to be preferred but care must be taken to ensure that the correct conformational form or oligomeric state is considered. If working with a modelled structure then one must always bear in mind that certain inaccuracies may be inherent in the modelling procedure. One can now proceed with mapping out the properties of the putative recognition sites.

Although it is clearly better to have a knowledge of the three-dimensional structure of the drug target, this information can be double-edged if one is unsure as to exactly where the ligand will bind, or which residues are important triggers for activity. For enzymes this problem is lessened as the active site quite often resides at the bottom of a clearly defined cleft in the structure, with some access to the surrounding solvent environment. But this is not always the case. If the binding site is unknown then the receptor mapping techniques outlined below must be utilized across the whole structure, with reference to mutation studies and any other available information.

Grid maps

If we take as our starting point the hypothetical situation where we know the structure of the receptor and the position of natural substrates or signalling

Goodford, P. J. (1985). *J. Med. Chem.*, **28**, 849–857.

molecules within that active site, two questions must be asked before designing a new ligand. If the ligand is to block the receptor (an antagonist or inhibitor), first, we must limit ourselves to thinking about which parts of the molecule could be replaced by alternative substituents without affecting the binding properties i.e., we are looking for potential bio-isosteres. Secondly, we must direct our attention to the remainder of the binding region to get an idea of where additional groups could be incorporated into the new ligands such that optimum binding is achieved. Fortunately both of these questions can be answered using the same approaches.

The simplest and most systematic method, which now forms the basis of many variants, is to investigate the energetics of interaction between a representative probe group and the protein. First, the target protein is surrounded by a grid which defines potential interaction sites. Next, a probe group – for example: water to represent –OH; ammonia for an amine – is placed at each site in turn and the interaction energy calculated. Since both the protein and probe groups are kept rigid the energy contains only intermolecular terms: electrostatic, van der Waals and hydrogen bond, but more complex formulations could be used. Many of the grid points can be eliminated as being 'inside' the protein, and hence will have large unfavourable energies, but the remaining points can be contoured according to their energy values allowing clear graphical interpretation of the results. These contoured regions can be selected to display only the most favourable sites (i.e., those that have a reasonable negative energy). Figure 6.4 shows such a map for an OH⁻ probe calculated in the active site of the enzyme dihydrofolate reductase.

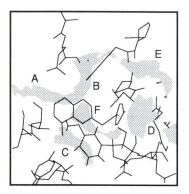

Fig. 6.4 Grid map of an active site.

The flexibility of such an approach is most clearly seen when probe maps from a number of different characteristic groups are combined and compared. In this way a very complete picture can be built up of 'what can go where' in the active site of the target molecule. From the relative position of acceptable hydrophobic binding areas and potential hydrogen bonding sites, say, something equivalent to a detailed pharmacophore can be generated. Additionally, one is not restricted to groups which are present in known binding agents; the suitability of any functional group can be tested simply by repeating the calculation with a set of appropriate non-bonded probe parameters. As implemented in most modelling software the method is more complex that outlined above. Directionality of functional groups can be accounted for by considering different orientations of the probe at each grid point, and additional energy functions can be used to accommodate environmental effects and ensure optimum hydrogen bond geometries. The validity of the method was clearly shown by its ability to reproduce the positions of ordered water molecules in x-ray structures, showing that these molecules must be considered as bound to the protein.

Multiple copy minimizations

Miranker, A. and Karplus, M. (1991). *Proteins: Structure, Function and Genetics*, **11**, 29–34.

A second method, which produces a similar end result, is based not on systematic searches on a grid but on energy minimization. Again typical functional groups are represented by small probe molecules, but these could easily be extended to include 'linker' atoms or groups. First, the active site of the target protein, or its complete surface, is surrounded by many hundreds of copies of the probe molecule. It is essential that these are randomly placed to eliminate bias due to the starting conditions. The next stage is to minimize all of these molecules simultaneously with respect to the protein's structure. In the minimization procedure the probe molecules cannot see each other, in the sense that no forces are calculated between them; only protein-probe interactions are considered. In this way very many trial orientations or starting positions can be tried at very little extra computational expense. If the minimization procedure is successful the resulting probe positions should show some clustering around favourable binding sites and each cluster can be reduced to a single representative molecule. These can then be used as starting configurations for more rigorous scans of the local environment. As in the previous case the most energetically favourable positions can be contoured for graphical display and the maps from different probe molecules combined. One very important extension which has been made to this method is to allow the inclusion of protein flexibility and so cooperative movements between a ligand and its receptors are simulated.

Receptor maps produced by the above methods are useful in that multiple maps can provide suitable alternatives to, say, the natural substrate's functionality. But they give almost no clues as to what kind of framework should be placed between these groups. At this stage some kind of 'docking' procedure must be applied to place putative ligands in the active site while satisfying the constraints of the allowed regions.

Docking methods

A considerable number of docking procedures exist in the literature. These range from the use of interactive graphics to manipulate the position of the ligand, to the completely automatic procedures which are becoming increasingly powerful for screening databases of molecules. One novel extension of the former was for the user to make adjustments to the molecule's position *via* a mechanical arm which provided resistance proportional to the forces felt by the ligand; an occasionally painful experience. However, tests showed that social science undergraduates were as successful as trained molecular modellers in finding the correct ligand position!

In principle, many docking algorithms follow a quite similar pattern. A typical flow chart is shown in Fig. 6.5. Usually the first stage is to represent the molecules by their solvent accessible surfaces. This smoothes out the exterior of the protein while at the same time defining the size of the molecules. Starting from a number of different relative orientations the two molecules can then be brought together. The driving force for this movement can simply be crude translation and rotation of one structure until there is contact, or alternatively one molecule can be moved subject to force field of the other. More sophisticated methods carry out their moves using a Monte Carlo algorithm to direct both rotations and translations. Rapid convergence to a minimum can be achieved by gradually cooling the simulation temperature as the molecule appears to be descending into a potential well. In effect the lower temperature slows down the extent of movement of the

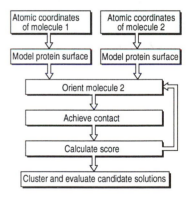

Fig. 6.5 Flow chart for typical docking algorithms.

ligand by allowing an ever decreasing number of successful moves. In all of these cases the intermolecular energy can be evaluated using a traditional van der Waals plus electrostatics function, but this will have a very steep repulsive penalty if there is any interpenetration of the ligand with its receptor. Increasingly, so called 'soft potentials' are being used which have much less drastic short-range repulsions and allow a certain amount of overlap. In this way they are thought to account for induced fit by the receptor after minimization of the complex. Once a sufficient number of different solutions have been obtained from a range of starting orientations, the complexes can be refined and clustered to remove degenerate structures. At this point it is advisable to include as many experimental data as are known to allow elimination of any structures which are clearly inconsistent. Having found a suitable binding site on the protein, other molecules can be docked simply by superposition. Alternatively, the user might calculate a receptor map to determine which other groups could be used at that site.

Another successful docking method works, in the first instance, with the volume of the receptor site. Molecules can be matched according to how their shape fits this volume and then scored using an appropriate energy function. As with most of the other methods the program treats the two molecules as rigid entities, and this is widely viewed as the greatest shortcoming of such methods. However, to include flexibility would increase the size of the problem considerably. Typically, it is implicitly understood that all of these methods provide starting geometries for energy refinement.

6.4 Generating new structures

Thus far we have concentrated on using ligands of known structure to provide information about the structure and properties of the binding site. The next step is to use this information to suggest new ligands with greater binding affinity.

Three-dimensional databases

Traditionally medicinal chemists would take as their starting point a lead molecule which was either one they had synthesized themselves or perhaps a competitor's structure. This 'lead' molecule would then be modified: different substituents could be added; ring systems would be modified, until an active and patentable molecule was found. However, the crucial feature of this process is that the new molecule is likely to look quite similar to the original; there has been no real leap to a different class of compound. Recently, computational methods have come to the chemist's assistance. This has come in the form of techniques to search vast databases of molecules.

All established pharmaceutical companies will have a huge library of structures which they have made, but often these putative drug molecules will only have been tested for a very limited range of activities. Perhaps other activities reside with these structures. The first stage is to convert the two-dimensional structural formulae into a database of three-dimensional structures. Automated procedures now exist which use knowledge-based approaches – they interpret rules for geometry and stereochemistry derived from known structures – to carry out this transformation rapidly. In most circumstances this will produce a database of tens of thousands of structures. This database can be further supplemented by the inclusion of other useful structural data such as ring centroids, likely protonation state at physiological pH, lone pair vectors and so on. The database can now be used in conjunction with information acquired from receptor mapping – that is to say, the pharmacophore – to locate new structures. Since the database contains information about the three-dimensional structure of the molecules, the type of questions asked are related to the three-dimensional structure of the receptor. A typical type of query is shown in Fig. 6.6, with the elements of the pharmacophore being the aromatic ring and the amine nitrogen; these require the spatial arrangement shown for optimum activity. This query is then tested against the complete database. Many of the solutions produced will be molecules known to be active at the particular receptor but it should be hoped that other, interesting structures which include this arrangement of the pharmacophore elements turn up. Searching databases simply on the basis of a pharmacophore still leaves the problem that the adjoining segments of the molecule are not defined so care must be taken to test the new structures against all known data on the receptor model, including allowed and disallowed volumes.

In an earlier section it was mentioned that projecting molecular properties onto a regular shaped geometric shape provides a useful, simplified representation of the molecule for molecular superpositions. Similarly, this method can also be applied to database screening. By extending the comparisons to a molecular property, longer range effects (modulation of the value of pK_a by distant electrostatic groups for example) can be accommodated into the comparison procedure. This has the additional advantage that the comparison is no longer being carried out on an atom-to-atom basis, providing much greater scope for topologically different molecules to be picked out. If a second screen, based upon molecular shape, is then carried out by the same procedure it is clear that one stands a reasonable chance of finding a molecule which would be a novel starting point for drug design. The second major advantage in the use of a regular shape is that fitting can be done very quickly using precalculated rotation matrices. Given that this approach seems at first glance to be fairly crude – reducing a molecule to at most twenty points on an extended surface – it has proved to be remarkably successful in locating useful molecules. It should also be noted that although the above discussion has been restricted to the situation where a single company's database of thousands of molecules is being searched, the complete compound index of *Chemical Abstracts* is being converted into three-dimensional form allowing millions of molecular shapes to be evaluated.

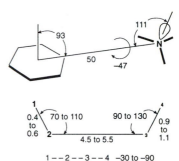

Fig. 6.6 A database search query based upon a typical CNS pharmacophore.

Spheres and icosahedra are used for database searches in van Geerestein, V. J., Perry, N. C., Grootenhuis, P. D. J., and Haasnoot, C. A. G. (1990). *Tetrahedron Computer Methodology*, **3**, 595–613.

De novo design methods

Another approach is to attempt to create, *de novo*, new molecular shapes which will fit a particular receptor site model. An ideal starting point for this approach would be a series of probe maps, as derived in a previous section.

This would define a set of suitable anchor points for particular functional groups. The next stage would be to assemble these into a single molecule using suitable linker regions, say, ring systems, such that the binding groups would be held quite rigidly in place. If the complete set of suitable linker groups is considered to be an ensemble of atomic sites, even more diverse ways of joining the functional groups may be devised by linking appropriate atom centres. The principal advantage of this approach is that one is no longer relying on existing molecules for information, as would be the case in a database search, therefore it is possible that completely novel entities will be found as lead compounds. Furthermore, if one is working with a three-dimensional structure of the receptor it will be immediately apparent if the linker pieces fit snugly into the binding site. Although much more difficult, it is also possible to use this type of method in conjunction with the active analog approach. Clearly, one will be limited by starting from a reduced set of functional-group binding sites, but, if the starting model is suitably well defined, the accessible and inaccessible volumes should be unambiguous.

Genetic algorithms

A further method for generating new structures has recently received considerable attention. This uses so-called 'genetic algorithms'. These programs allow the starting molecule to be modified over successive generations of structures in an analogous way to the random changes in DNA during replication. Changes take place in two different ways: crossover and mutation. In crossover, features of two molecules are combined to produce two new molecules (offspring), each of which contains half of each parent molecule. The second genetic process is mutation: that is to say, single features of molecules are allowed to change between successive generations, subject to an allowed mutation rate. Typically this would take the form of changing one functional group for another. The control process which determines which molecules proceed to the next generation is selection; again analogous to the Darwinian ideal. Molecules can be tested for 'fitness' – their goodness of fit to a proposed receptor model – with only the most likely structures being propagated forward. Alternatively, selection can be based almost entirely on chance, which would allow less useful structures to continue but if further generations produced no better structure this avenue would be discontinued. The main attraction of this approach is that it is completely random, and while this may produce ridiculous structures much of the time it can also turn up interesting combinations. Apart from setting mutation rates, the only other control which the user has is over the set of functional groups and structural units which are incorporated into the 'gene pool'.

New structures can be found by a number of methods and many potentially novel compounds can be 'invented' by the various computer algorithms. But the likelihood of these structures being useful is highly dependent on the receptor model used initially.

The last question to be addressed is how similar are two structures, and can this be quantified?

6.5 Molecular similarity

One of the most difficult comparisons that a medicinal chemist is faced with is to consider how similar two molecules actually are. This is clearly a rather loose idea (just as the word similar implies) however considerable inroads have been made towards providing useful indices of similarity. To complicate things further one must also consider similarity of more than one property for the method to be a useful tool for ligand design; molecules must show both electrostatic and shape similarity.

In the preceding sections we have taken it for granted that sets of molecules can be superimposed without difficulty, or that the comparison of new molecules is routine. Clearly this is only so if the molecules are based upon related chemical frameworks, but if one is pursuing novelty then this will not be the case. The idea of shape similarity is obvious and stems from some of the earliest analogies about the nature of interactions between big molecules and small molecules: the lock and key. Electrostatic similarity is less easy to picture in that it is a property that results not just from an individual grouping but also from the influence of the rest of the molecule on that site. Remembering that the electrostatic potential at a point results from the sum over all charges, it should be obvious that different charge distributions could result in not dissimilar potentials.

For further details see Richards, W. G. and Hodgkin, E. E. (1988). *Chem. Brit.*, 1141–1144, and Good, A. C., So, S.-S., and Richards, W. G. (1993). *J. Med. Chem.*, **36**, 433–438.

The earliest attempts to compare molecules concentrated on similarity of electron density; this property being readily obtainable from molecular orbital calculations. One of the first formulae was

$$R_{AB} = \frac{\int \rho_A \rho_B \, dv}{\left(\int \rho_A^2 \, dv \right)^{\frac{1}{2}} \left(\int \rho_B^2 \, dv \right)^{\frac{1}{2}}} \tag{6.2}$$

where ρ_A and ρ_B are the electron densities of molecules A and B respectively, and these are integrated over all space, v. The denominator normalizes the function to give a similarity value (R_{AB}) between 0 and 1. The principal problem with this formula is that it compares shape of charge distribution, not magnitude, thereby reducing its utility. An alternative formulation due to Hodgkin and Richards is

$$R_{AB} = \frac{2 \int \rho_A \rho_B \, dv}{\int \rho_A^2 \, dv + \int \rho_B^2 \, dv} \tag{6.3}$$

which compares both magnitude and shape.

The major problem with these methods is the complexity of the integrals involved and their computing requirements. This has necessitated the search for other parameters to compare. One such alternative is the electrostatic potential which is a much better indicator of a molecule's electrostatic properties than the atom-centred charges, while at the same time it can be rapidly calculated from the molecular wavefunction or from a point charge distribution. A practical problem is that the electrostatic potential goes to infinity as it approaches the atom centres but this can be overcome by considering only the potential beyond the van der Waals surface of the molecule. Thus the electrostatic potential can be calculated on a grid surrounding the molecule and the similarity of potential can be calculated using Eqn. 6.3. If one uses point charges to calculate the potential then conformational flexibility can be introduced to find the fit between two molecules which maximizes their similarity; a tool of considerable importance for many of the methods used in this chapter.

Similarity of molecular shape is also quantifiable in a very simple way. First, a pair of superimposed molecules is surrounded by a rectilinear grid. Each grid point is then tested as to whether it lies inside the van der Waals surface of one, two, or neither of the molecules. Similarity is then evaluated using Eqn. 6.4.

$$S_{12} = \frac{B}{\left(T_1 T_2\right)^{\frac{1}{2}}} \tag{6.4}$$

where B is the number of points common to both structures and T_1 and T_2 are the total number of points falling within the surfaces of molecules 1 and 2 respectively (Fig. 6.7). One interesting corollary to shape similarity is *dissimilarity* (*D*) which is defined as 1 - *similarity*. This has recently been shown to have considerable use in comparing chiral molecules and allowing the rational explanation of previously inexplicable structure-activity relationships.

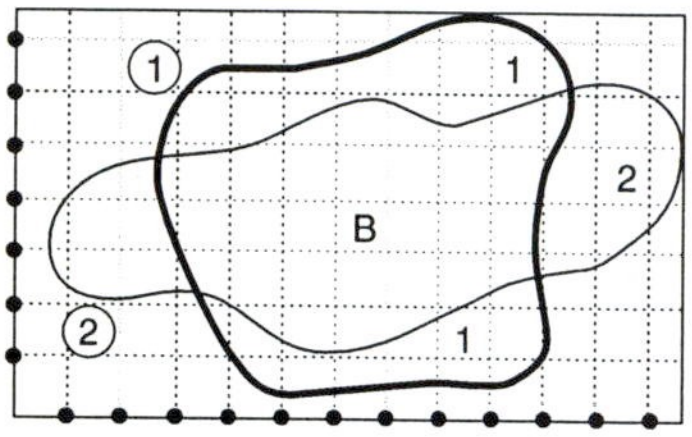

Fig. 6.7 Similarity of shape.

6.6 Summary

A vast array of methods is now available to facilitate the design of novel ligands for biomolecules, particularly putative drug molecules. Often the target for the drug is not structurally defined and one must infer structural information from the molecules which do and do not show activity in biochemical assays. By superimposing molecules one can define the most likely binding units (the pharmacophore) and the properties and volumes of other sites of the receptor. If the receptor structure is known it can be mapped to show possible binding sites for particular functional groups. This can be treated as the equivalent of a pharmacophore for the purposes of database searches or as constraints for *de novo* ligand design. In the latter a computer program attempts to fit molecular fragments between functional group binding sites to create new molecules, or alternatively, genetic algorithms can be used. The final test for any new molecule is whether it resembles the existing active molecules. Molecular similarity techniques allow both shape and electrostatic similarity to be quantified. These techniques can also be useful at the earliest stages in the design process where optimal superposition of molecules is required.

Concluding remarks

The object of the preceding chapters was to provide an introduction to many of the methods currently used by practising computational chemists; in no way was it intended as a 'how to do it' manual. Although the emphasis has been on computer modelling of biological systems, many of the methods are equally valid in other areas of research. One can imagine mapping out a cavity within a zeolite in much the same way as we have described for enzymes; the conformations of a polymer can be investigated by Monte Carlo or molecular dynamics methods; solid state systems can be treated as periodic ensembles of molecules, but subject to particular symmetry constraints.

That computational chemistry now plays a role in such a diverse range of chemical disciplines is a clear sign of its maturity as a field. Chemistry should be concerned with molecular explanations and it is pictures of this type which computational chemistry provides. As long ago as 1959 it was noted by Charles Coulson, in an after dinner speech, that there appeared to be two cultures within quantum chemistry. The first was concerned with obtaining ever more accurate solutions for very small systems while the second was prepared to sacrifice absolute accuracy and work with more approximate methods in the hope of gaining insight into 'real' chemical problems: molecular pictures. The latter are now computational rather than theoretical chemists. This distinction also serves as a motivation for understanding the underlying principles of the subject; before placing an interpretation on the results of a calculation one must be fully aware as to whether they are valid within the limits of the level of theory.

Finally, we must look to the future. The most obvious impact will be technological; faster and more powerful computers should enable much larger systems to be investigated. Also, the trend towards parallel processing will lead to a re-evaluation of many algorithms in the hope that they can be made to run more efficiently. The second major development should be in the computational methods themselves, in particular the energy calculations. Vastly increased computer power should allow the utilization of less approximate force fields or even large scale quantum chemical calculations. In either case the developments will have obvious benefits in both academic and industrial research.

Further reading

General

Allen, M. P. and Tildesley, D. J. (1987). *Computer simulation of liquids.* Clarendon Press, Oxford.

Atkins, P. W. (1990). *Physical chemistry*, (4th edn). OUP, Oxford.

Branden, C. and Tooze, J. (1991). *Introduction to protein structure.* Garland, New York.

Burkert, U. and Allinger, N. L. (1982). *Molecular mechanics.* ACS Monograph no. 177, American Chemical Society, Washington.

Cartwright, H. M. (1993). *Applications of artificial intelligence in chemistry.* OUP, Oxford.

Chandler, D. (1987). *Introduction to modern statistical mechanics*, chapter 6. OUP, New York.

Clark, T. A. (1985). *A handbook of computational chemistry.* Wiley, New York.

Richards, W. G. and Cooper, D. L. (1983). *Ab initio molecular orbital calculations for chemists*, (2nd edn). Clarendon Press, Oxford.

Stewart, J. J. P. (1990). *MOPAC: A semi-empirical molecular orbital program.* ESCOM, Leiden.

van Gunsteren, W. F. and Berendsen, H. J. C. (1990). *Angew. Chem., Int. Ed. Engl.*, **29**, 992–1023.

White, D. N. J. (1978). *Molecular Structure by Diffraction Methods*, **6**, 38–62.

Other applications

Beveridge, D. L., Swaminathan, S., Ravishanker, G., Withka, J. M., Srinivasan, J., Prevost, C. *et al.* (1993). In *Topics in molecular and structural biology Vol. 17, Water and biological molecules,* (ed. Westhof, E.) pp 165–225. Macmillan, London. (DNA).

Catlow, C. R. A., (ed.) (1992). *Modelling of structure and reactivity in zeolites.* Academic Press, London.

Colbourn, E. A., (ed.) (1994). *Computer simulation of polymers.* Longman, Harlow.

French, A. D. and Brady, J. W., (eds.) (1990). *Computer modelling of carbohydrate molecules.* ACS Symposium Series, American Chemical Society, Washington.

Pastor, R. W. (1994). *Current Opinion in Structural Biology*, **4**, 486–492. (lipids).